W9-BIS-701

Also by Ruth Moore

MAN, TIME, AND FOSSILS (1953)
CHARLES DARWIN (1955)
THE EARTH WE LIVE ON (1956)
THE COIL OF LIFE (1961)
NIELS BOHR (1966)

These are Borzoi Books
Published in New York by Alfred A. Knopf

MAN IN THE ENVIRONMENT

MAN
IN THE
ENVIRONMENT

Ruth Moore

With the cooperation of the Field Museum

Illustrated by Marion Pahl

ALFRED A. KNOPF *New York 1975*

THIS IS A BORZOI BOOK
PUBLISHED BY ALFRED A. KNOPF, INC.

*Library of Congress Cataloging in Publication Data
Moore, Ruth E.
Man in the environment.
Includes index.
1. Human ecology. I. Title.
GF47.M6 1975   301.31   74–21282
ISBN 0–394–49565–9
ISBN 0–394–73046–1 pbk.*

Manufactured in the United States of America

First Edition

CONTENTS

Acknowledgments

The fate of this earth on which we live is certainly the ultimate problem before us. And most disturbingly, that fate—the continuance of the earth as a livable planet for man—can no longer be taken for granted. Pollution, overcrowding, and depletion have put the future into question. It has been recognized that if the full scope and dimensions of the earth's present predicament are to be understood, and if preventative action is to be taken in time, a part of the responsibility for creating this understanding and awareness must rest with museums of natural history. They are uniquely qualified to see and to demonstrate, for the earth and its inseparable life are their special province.

An exhibit that will bring the story of the earth and its total relatedness and vulnerability before the people of the United States has been planned by the Field Museum of Natural History of Chicago. This book was undertaken to organize these immensely complex concepts and to provide a base for the development of the exhibit—Man in His Environment.

To make the concept explicit in a book and visible in an exhibit—to show people what they must know—was an overwhelming task. It has rarely been attempted as a whole, and it could be approached now only with the full cooperation and scientific knowledge of many scientists at the Museum and many outside it.

For general formulation of the book and for constant assistance I wish to thank Robert F. Inger, Assistant Director of Science and Education at the Field Museum, and Henry S. Dybas, Curator of Insects.

Other members of the Museum staff gave continuing direction and assistance in their own fields of specialization. Without attempting to cite all, I should like particularly to thank Glen Cole, Associate Curator of Pre-History; Matthew Nitecki, Associate Curator of Fossil Invertebrates; Edward Olsen, Curator of Mineralogy; Bertram Woodland, Curator of Igneous Petrology; and Rainer Zangerl, Curator Emeritus, Department of Geology.

Without the interest and encouragement of E. Leland Webber, Director of the Field Museum, the book, of course, would have been impossible.

All of us were aided in addition by a group of consultants whose ideas and criticisms were invaluable. I wish to express our thanks to: Rabbi Moshe Davidowitz, New York University; Edward S. Deevey, Jr., Professor, University of Florida; Edward T. Hall, Northwestern University, Chicago; Charles H. Long, Professor, Divinity School, University of Chicago; Edward A. Maser, Professor, Department of Art and the College, University of Chicago; Herbert J. Muller, Department of English, Indiana University; Harvey S. Perloff, Dean, School of Architecture and Urban Planning, University of California, Los Angeles; and Cecelia Tichi, Depart- of English, Boston University.

The exhibit is scheduled to open at the Field Museum in September 1975. It also will be shown at other museums throughout the country.

Ruth Moore
Chicago, December 1, 1974

MAN IN THE
ENVIRONMENT

Chapter 1

AT THE START:
A FOREWORD

The azure sphere with its wreathing of white clouds—our earth—revolves in a lonely, separate orbit. All around stretches nearly illimitable space. Even the moon, the nearest solid body, is 238,857 miles and days of space travel away. This earth, with its thin, translucent film of atmosphere, must then be self-contained.

It is true, of course, that the sun pours its rays down upon us and thereby gives us energy, warmth, and light. But the only additions that man ordinarily considers solid, a rain of fine dust from outer space and a few chunks of iron and stone—meteorites—are negligible. They add little to the earth's substance.

Essentially, all that we have is here now, and it has been since it was assembled at the time of the planet's formation. From the beginning, the use and reuse, or cycling and recycling, of this original material have given us the air we breathe and the atmosphere that protects us from incineration by the sun; the oceans and our other waters; our rocks, minerals, and other physical resources; and the corpus of all life, plant and animal, including, of course, the approximately 3.5 billion humans of today.

Until the last few centuries men and nature worked well together, or at least nature was allowed to function without consequential interference by man. The earth seemed large and its resources inexhaustible, and for those living in earlier times this was undoubtedly true. Few even questioned that man, if he had the capacities, could find enough resources for his needs, or that any actions of man could materially affect the earth itself.

Now men have become numerous enough to overcrowd many sections of the earth and the earth itself. Our industries, agriculture, mining, and cities have expanded until their outputs and wastes are polluting the long unpolluted land, air, and water, and are rapidly depleting the nonrenewable earth resources on which all depend. Even more disturbingly, the combined overcrowding, pollution, and depletion are threatening the natural cycles and the huge diversity of natural things that sustain all that exists on the earth, under it, or closely above it. The problem is new, for in the past it did not exist, but now for the first time the earth is in danger.

Some scientific extrapolations go further. They logically show that if the present exponential expansion continues on this finite earth the result must be catastrophe—starvation and collapse of man and civilization. The planet in such an event could become uninhabitable.

But this does not have to happen, and it cannot be permitted to happen. The exponential expansion that inevitably would overrun and ruin is not itself inevitable. New studies, using the advanced techniques of the computer, are developing ways out, and although this work is only in its earliest stages, and no one can say with certainty how livable limitation can be achieved, the principles are becoming clear. Realistic appraisals are also being made of some of the foreseeable problems and costs, economic, social, and political, that must be faced if change is to be made. In addition, understanding is growing of the ecological problems of the earth. It is also becoming evident that in the end all will depend upon us, our conscience and understanding. Will we comprehend the dangers in time, and will we undertake the changes necessary to maintain a viable earth?

It will be the purpose of this book to examine the earth as it really is—that is, as a natural unit, a functioning, evolving whole of which all living and inanimate things are interacting parts. It is rarely considered in this way. But only if we look at the earth and ourselves as one can we judge what is happening to the earth and to ourselves and what must be done.

In this attempt to picture the earth in its historic, natural context, the book will consider the determinative formation of the earth, the evolution of its life, the workings of the earth, man's impact on the earth and on man, as well as the imperatives and the alternatives open to us and the implications—some of the formidable changes that may be required if life on this planet is to endure.

An associated exhibit at the Field Museum in Chicago will illustrate the main and all-important point—that man is a part of nature, not its master, and that nature is a part of man. On this concept our survival rests.

Chapter 2

ALL THAT WE HAVE

Any understanding of
the predicament of the earth and what can be done about it
must begin with an understanding of what we have—the earth
itself, the base of it all.

As the saying goes, it is all we have. Any likelihood that
resources can be brought in from other planets is as remote as
the great distances to those planets.

Exactly what the earth's beginning was no one can say
with any certainty. In the eighth century B.C. Hesiod, tending
his flocks on the slopes of Greece's holy Mount Helicon,
heard the muses whisper that in the beginning there had
been "chaos, dark, wasteful and wild" and that into this void
"wide-bosomed earth, the ever sure foundation of all" was
born. Some thirteen centuries later the Roman Lucretius
firmly rejected the earlier myths and argued that "multitudinous atoms," swept along by multitudinous clashes and
their own weight, had come together and formed the earth.

Later still, in the nineteenth century, Laplace, the
French mathematician-astronomer, suggested that the earth
might have originated in a huge cloud of blazing gas thrown

off by the sun. If that had been the case, it was later determined, the superheated gases would have liquefied into a
molten orb, with a composition quite unlike that of the earth.

In the first decades of the twentieth century Thomas
Chrowder Chamberlin of the University of Chicago proposed
that the earth may have been assembled when some of the
not uncommon "knots of matter" in some of the nebulae
began to gather in other pieces of matter and cosmic dusts.
Many changes have been made in Chamberlin's "planetesimal
hypothesis," but many astronomers hold today that the earth
and the other planets of our galaxy began as swirling eddies
and clusters of cosmic dust and gas. Indeed, modern tele-
scopes reveal the existence today of many assemblies of the
kind that might give birth to a star, or an earth.

According to this theory, as the eddy that was to become
the earth followed its endless orbit around the sun, and as
this mass rotated, it began to condense and consolidate.
Chemical and physical examinations of the rocks and structures of the earth show that they are what would be expected
if their assembly had been cool. They do not exhibit the ob-
vious effects of having been subjected to high heats. It is
concluded, therefore, that the initial formation of the earth
occurred at low temperatures.

However, among the elements in-gathered by the young,
forming earth were some that were radioactive. As all radioactive substances do, they constantly decayed or steadily lost
their radioactivity in the form of heat, thus creating heated
areas. In some cases enough heat was generated to melt some
of the original silicates and to reduce some of the iron oxides
to iron. By some 4 billion years ago the original mélange had
differentiated into a molten iron core, the thick mantle, and
the thin outer crust that persist to this day.

The earth then had its essential substance and organization. As solid and unyielding as it may seem to its present
occupants, the earth was far from uniform in composition
and far from immobile. The lighter parts, buoyed up on the

denser masses, may have been pushed and shoved together to form the first land masses.

In a coolly assembling earth, any water created by the chemical processes and the pressures would have combined with the original silicate rocks to form hydrated silicates—which are exactly the rocks that are found in quantity in the earth's crust. Later meltings and movements may have released some of these waters.

That such a process occurs is well demonstrated in the Snake River area of Idaho. There, a huge dome of melted rock that pushed to the surface some millions of years ago is still discharging hot water at more than one hundred known hot springs. William W. Rubey, a former president of the Geological Society of America, calculated that in 3 billion years the discharge of the world's hot springs could produce a volume of water one hundred times that of the present oceans. Other "juvenile," or new, water would also have been erupting from the world's volcanoes.

Rubey also demonstrated that the crystallization of a shell of water-holding rocks, like those of the crust, could have accounted for all the water of the oceans. Another geologist has estimated that the oceans could have derived from the escape of 5.8 percent of the water in the sialic rocks of the crust.

In one such way or another, or perhaps by a combination of all such water production, the earth obtained its waters from its own substance and its own interiors. There was and is no other possible source. No rain falls on us from outer space, from outside our own atmosphere. No space vehicle traveling outside our atmosphere ever runs into a rainstorm.

Our oceans, therefore, were filled and replenished by natural processes that man with all his technology can neither speed nor slow. In any short period of time, such as the last few thousand years, the earth's total water supply is little altered. Modern man must for the most part live with the water he has. Life is too short to count on a large increase in supply from the earth, our only source of new, "juvenile" water.

If the oceans and other waters become seriously polluted, man will have to live—or die—with them, for not enough added new water will be forthcoming from springs and volcanoes to meet modern water requirements and to overcome pollution. It must be recognized that the waters of the earth are a limited, little expandable resource, which is one of the first principles of the earth, and one that must constantly be borne in mind as man fights pollution and the possible ruin of the earth.

At first no life stirred in the accumulating new waters or on the lighter land masses the waters surrounded. Exactly how life originated in these waters, no one absolutely knows, although several possible clues have been developed by scientists. One theory is that some of the lavas pouring out from the volcanoes carried some carbon compounds with them. It is, in any case, certain that in some of the volcanic vents carbon was fused into diamonds and graphite. So carbon must have been there, and some of the carbon that reached the surface and its waters may have come together in tiny chains and globules. The little strings and blobs may have been something like similar carbon structures found in fossil oil and pollen; such carbon structures are considered organic, or characteristic of life.

Were volcanic eruptions, and the lavas, the source of the carbon that is the foundation of all life? Or could this carbon of the earth's interior have come from outer space, with the original dusts and debris that formed the earth itself? Among the reasons for hypothesizing that it may have are the carbon compounds in stony meteorites and planetesimals that have crashed into the earth.

From time to time reports have been made of the finding of carbonaceous materials in this one type of meteorite. Until recently, however, it was assumed that the presence of the carbon compounds had to be attributed to contamination.

The carbon could have been picked up as the meteorites fell through the atmosphere or in their contact with the earth's surface.

In 1969 other stony meteorites fell in southern Australia. This time Cyril A. Ponnamperuma, and his associates at the Ames Research Center, made the most rigorous of tests, and they were able to rule out contamination. The carbonaceous materials in the Australian meteorites were, by all indications, original. Most significantly, some of the amino acids—the carbonaceous material—had never before been seen in terrestrial organisms. The conclusion was that the amino acids had been formed in outer space in the parent body from which the meteoritic fragments had come. As the fragments orbited in space, the amino acids they contained would have been unchanged.

Perhaps other debris from other broken-up planetary bodies carried other carbon compounds into the clusters that consolidated into the earth. If so, carbon was brought to the earth in its earliest stages.

Investigation has disclosed another possible origin for the carbonaceous materials. As the earth consolidated, some of its gases may have enveloped the young globe, much as does our present atmosphere. Sunlight streaming through this cloudy covering and the flash of lightning or other reactions may have converted some of the gaseous carbon into carbon chains or globules.

Laboratory experiment has demonstrated that such a reaction can occur. Stanley L. Miller, when he was a graduate student working with Harold C. Urey at the University of Chicago, passed an electrical discharge through a mixture of gases approximating the gases of the early earth. The mixture soon turned pink, and after one week it was red: carbon chains, including precellular bodies and amino acids, had been formed. Later work by other investigators has shown the same productions when similar gases are treated with a variety of agents, including ultraviolet rays, X rays, gamma rays, and heat.

Perhaps carbonaceous materials brought toward the surface of the earth by volcanoes accumulated in underground pools or in the damp crevices of rocks. Perhaps amino acids brought in by meteorites or falling from the atmosphere came together along the edges of the earth's new waters. In time, however, carbonaceous materials filled the seas, producing a rich "broth" or "soup."

By 3 billion years ago the waters were swarming with microscopic rod-shaped organisms and tiny spheroidal bodies. The carbon strings and globules had become living things—and at this point it is no longer necessary to speculate, for their fossil remains have been found and photographed in unmistakable reality.

Elso S. Barghoorn, professor of botany at Harvard University, discovered the dawn organisms in a high cliff of stratified rock in the gold-mining district of Africa, embedded in black, gray, and greenish cherts. In places the chert beds, known as the Fig Tree formations, were four hundred feet thick. The stone had been formed under water, and the algae and bacterialike organisms drifting down through the waters had come quietly to their resting place, and had remained there, essentially undisturbed, for more than 3 billion years. (The rocks were dated by radioactive measurements.)

At first Barghoorn studied the microorganic material under the light microscope. It was difficult, though, to see distinct bodies. With the assistance of J. William Schopf, sections of the organism chert were then "shadowed" with metal for examination under the electron microscope.

Suddenly the rodlike structures stood forth. They were exceedingly small, a mere 0.5 to 0.7 microns in length, but they could be seen clearly both in profile and in cross section. In cross section, not only were the cell walls visible, but an inner and outer layer could be seen. The Fig Tree beds also held threadlike, microscopic filaments that were almost certainly of biological origin, and which closely resembled decomposed plant material. On later tests tiny spheroids were discovered.

"They may be among the evolutionary precursors of the modern blue-green algae of the coccoid group," said Barghoorn.

The rods were named *Eobacterium isolatum,* to denote dawn bacteria growing separately rather than in colonies. They were a new genus and species—a high distinction. The spheroids were called *Archaeosphaeroides barbertonensis,* or ancient globules found near the town of Barberton.

"The existence of these two organisms, successful inhabitants of an aquatic environment more than three billion years ago, is evidence that the first evolutionary threshold— the transition from chemical evolution to organic evolution— had been crossed at some earlier date," said Barghoorn in a report in *Scientific American.* "We know that at least two living organisms appeared well before the first third of earth history had passed."

The carbon materials of the earth had somehow come together in minute living things.

The dawn organisms may have passed another determinative, world-shaping threshold. The diminutive globules may well have used the energy of the sun, which was still beating down uninhibitedly on the waters in which they lived, to manufacture food for themselves. If so, they took in carbon dioxide and gave off free oxygen. Until the Fig Tree era all the evidence from the rocks suggests the absence of oxygen and thus of an atmosphere as we know it. But some of the rocks laid down at Fig Tree show that small amounts of oxygen may have appeared. It could have come only from photosynthesis, or from the dawn plants' conversion of the sun's energy.

"The appearance of free oxygen was an event destined to have profound influence, both biological and geological, on the subsequent history of the earth," said Barghoorn.

A billion years later, or about 2 billion years ago, there

was virtually no question that plants were producing oxygen. In exposed beds of black chert, three to nine inches thick and extending for about 115 miles along the shores of Lake Superior and Gunflint Lake in Ontario, Canada, Barghoorn and Stanley A. Tyler, of the University of Wisconsin, found eight genera and twelve species of ancient algae. Some were filamentous, some spheroidal, some exquisitely star-shaped, and some umbrellalike. Still another species consisted of two concentric spheres. Although these species have long since become extinct, the Gunflint algae closely resembled modern photosynthetic blue-green algae, and some of their structures were closely like those constructed today by dome-building species. The Gunflint algae were true living plants. Unquestionably, they took in carbon dioxide and released oxygen.

The earth was on its way to acquiring its atmosphere. The oxygen given off by the dawn plants created both the air that would support all the oxygen-breathing life that was to come and the ozone that would shield that life from the burning ultraviolet radiation of the sun. An earthscape rather than a moonscape was becoming possible.

The envelope of air forming around the earth was produced, and would be renewed, strictly through the action of the plants with the sun. It was not some emanation or residue from outer space. The existence of the air and the layered atmosphere, therefore, depend upon the existence of the plants. If the plants of the earth were to be seriously reduced or devastated by any means whatsoever, man would have to know that his air and shielding from the sun would suffer. Though no one knows at what point serious damage might occur or how much greenery is necessary, the interdependency of plants and air is a certainty. Our air was created by plants and has been sustained by them for a billion years or more. And only in the last few centuries has there been substantial interference with this long-functioning process.

The plants of the Gunflint formation carried the evolution of the earth another long step forward. They had evolved a variety of forms, each distinctive and each particularly suited to its own environment. If some perished through some change in their own special environment, the chances were that others in other areas would survive, since not every place would become unfavorable at one time. Assurance was thus supplied by diversity. Life, as Barghoorn pointed out, had crossed a second evolutionary threshold—the threshold of diversification.

In a continuation of this series of discoveries, Barghoorn and his associates found evidence of the crossing of a third threshold. Another stratum of black chert, this one in the Northern Territory of Australia and in its Bitter Springs formation, yielded three groups of primitive green algae. These had lived in the shallow seas that covered this part of Australia about 1 billion years ago.

The green algae, unlike the earlier blue-greens, were capable of sexual reproduction. Each cell possessed a nucleus, instead of the scattered genetic material of their predecessors. A remarkable piece of luck left no doubt that this new state in the evolution of the world and its tenants had been reached, for the algae had been preserved in all the major sequences of sexual or mitotic cell division. The entire process could be seen as clearly as though it were happening under a scientist's microscope. Among the fossils were many single-walled cells, with the nucleus visible as a tiny round spot within the cell wall enclosure. In others the cell had elongated and a divided nucleus had drifted toward the two ends. In others a new cell wall had grown between the two halves, and in still others the two halves were just pulling apart. One cell had produced two, and the two were as complete individuals as the one had been.

The reassortment and recombination of the genetic materials in the offspring also produced difference. There was a varied assortment of the characteristics of the parent cells, and from this time on no two sexually produced individuals would ever again be exactly the same.[1] The difference supplied a base for natural selection; one offspring cell might be better adapted to the environment than the other and would be the one more likely to survive. Without the development of variability and individual difference, life might never have gone beyond the blue-green algae stage and the earth might have been populated solely by the algal slime. Such an outcome was conceivable, Barghoorn mused. In fact the algae and their precursors were the sole living occupants of the earth for nearly three-fourths of its history.

With the evolution of sexual division, natural selection could choose and the pace of evolution was speeded. It had taken 3 billion years to go from the first dawn organisms to the green algae. All the rest, the multitude of species and individuals, would come in about 1 billion years. The introduction of variability brought an "explosive" multiplication of species, which the fossil record traces.

By 600 million years ago the seas were alive with thousands of species of aquatic plants and animals. All animals were invertebrates, that is, without internal skeletons. They lived mostly in the shallows of what was then a universal ocean washing the shore of a single land mass, which is now called Gondwana Land.

This huge land mass, encompassing as it did all the present separate continents, still lay bleak, barren, and silent except for the storm and upheaval of nature. But it was becoming a more beneficent place. Part of the oxygen given off by the plants of the seas had formed a layer of ozone—a special form of oxygen—in the atmosphere and was tempering the sun's heat. The land lay open and, perhaps it could be said, welcoming. At least it was not impossible for life. The aquatic plants began to move inland from the shores, and

---

[1] Identical twins are the sole exception, and even they may show slight differences. Simple, nonsexual cell division, on the other hand, produces duplicates.

gradually the land's barrenness was covered by a mantle of green. With only air, water, the earth's minerals, and the tempered warmth of the sun the whole aspect of the earth was changed.

All the while, life in the sea continued to evolve. The rocks laid down about 425 million years ago contain the fossilized remains of a new kind of creature, one that looked something like a crudely formed fish. Instead of the soft, amorphous form of some sea life or the shells of clams, snails, and similar creatures, it had an internal skeleton and was armored with bony scales. And as it swam along the bottom it sucked up its food.

Some 60 million more years went by before some fish living in the shallows or in pools along the shores evolved fins that could be used as legs and gills that could function as lungs. They, like the plants before them, made their way out onto the land; some ancient deposits in Greenland have yielded some of their fossil remains. The crossopterygians, as they are called, had both lungs and well-developed legs as well as the traditional fish tail. They could crawl along the swampy shores and snap up the insects that were beginning to swarm over such areas. The crawlers were the amphibians, and the land beckoned them on. No others were there to contest their way.

Well beyond the beaches and estuaries, mosses, ferns, and seeding plants were flourishing, for the plants, too, had been evolving in their land domain. But though the lush vegetation offered plenty of food, the amphibians could not go far from the water. They always had to return to the shores to lay their soft, jelly-coated eggs. In time, however, some amphibians began to lay eggs encased in tough, leathery shells, which were internally fertilized and deposited in a safe, dry place to await the hatching of the young. The egg layers had won their freedom from the seas and the shores. They could live inland—and so the age of reptiles began.

In the dense, green jungles of the period, some of the

reptiles evolved into the largest animals ever to frequent the earth—the dinosaurs. Others—the pterodactyls—took somewhat lumberingly to the air. Some returned to the water, to give rise in time to whales and porpoises.

Except for the insects and the flying reptiles, the air was largely unoccupied, and it lay open to any venturer. Quite independently the birds evolved, not from the pterodactyls, but from the archsaurs, the same group that had fathered the flying reptiles and others. One of the feathered but still reptile-like birds fell, dying, into the calm, clear waters of a coral lagoon in the area the modern world calls Germany, and as the fine lime of the lagoon settled down upon the body, a nearly perfect imprint was made of it. When the cast was discovered in the 1950s the feathers were etched with the detailed exactitude of a fine bronze portrait medallion. If they had not been so precisely recorded, few would have classified the creature with the long neck, teeth, and strong hind legs as anything other than a reptile. The feathered link, which was named "archaeopteryx," was a primitive bird. Once again one group had given rise to another. The close relationship, however visibly concealed, was there.

While the reptiles still held sway over the land, one small member of the order developed somewhat longer and stouter legs. It also had warm blood, as did the birds, and a thick coat of hair that gave protection from the heat or cold. Some of the first may have suckled their young, even though they continued to be egg layers. The change in nurture was a critical one. The mammals, as these animals were named, produced fewer offspring, but the young were assured of food and of more protection than had been given any young before them. Survival became not a matter of chance, as when thousands of eggs floated free in the water, but of genetic fitness to survive. Thus the few could outdistance the many.

The new mammals, however, for all of their advantage, did not quickly take over the earth. At first they were few and furtive. But climates were changing; the earth was becoming cooler. The reptiles began to decline and the mammals started a phenomenal branching out, spreading into almost every part of the earth—the hot lands and the cold, the air and the water, the wetlands and the deserts. In each area they multiplied with remarkable success. Each was different, but the linkage to the line and to their predecessors was clear.

Large changes were occurring in this still early world. North and South America were pulling ever farther away from the African land mass, and the Atlantic Ocean was filling the rift that was thus created. The fauna and flora rode along with the land they lived on, and as the separation became wider and conditions changed, the riders on the two land masses evolved in their own way, as the ancestors they had left behind in Africa did in theirs. Each became different, though alike, too.

By 25 million years ago a new primate, a tailed monkey, lived in the New World. In the Old—Africa—another distinctive primate evolved, a larger, stronger-shouldered, tailless animal. The latter, an ape, swung arm over arm through the trees, only occasionally coming to the ground to move from grove to grove. On the ground the apes moved in a partly upright position by resting their forward weight on their bent knuckles.

As the millennia went by, some of the apes evolved a different jaw and teeth. The teeth were smaller and the face may have been shortened, in something of the way that was later to be seen in man. Louis S. B. Leakey found such fossil remains near Fort Ternan, Kenya—they were dated at 14 million years. Similar teeth and jaws were unearthed in the Siwalik Hills of India. Apes changing in such a way must have ranged widely through the continuous forests of Africa and Asia; the two continents were then joined. Although few of the fossilized bones have been found, such populations of apes may have reached into the millions, if there were even twenty to a square mile.

Despite the relative dearth of fossils, a new kind of evidence is showing both how the apes were evolving and how the genetic material—DNA, deoxyribonucleic acid—was passed along from one line to another. The evidence is from biochemistry, and it is precise.

From the beginning, from the first primordial broth with its microscopic rods and spheres, all evolution, all life, has been an up-building of DNA. The first coils of DNA—it is shaped like a spiral staircase—would have been a staircase with only a few steps. Life grew more complex as the coils, or staircases, lengthened and sometimes were altered at one step or another by mutation. The coils of the higher animals, and ultimately of man, came to contain more than a billion steps—and the change from the few to the billions was the change from bacteria to man.

At this stage it was apparent that if the DNA of various species could be compared step by step—if it could be seen where the DNAs were alike and where they differed—the whole record of evolution could be read. It would be possible to determine conclusively, rather than roughly, what species had sprung from what species, and in general when the separation had occurred. And although this still cannot be done, scientists have succeeded in deciphering the lineup of a number of protiens, and the protien lineup duplicates that of its respective section of DNA.

Indirectly, some of these lineups are readable. Studies disclosed that there are 141 amino acids or units in one chain of hemoglobin, the red substance of the blood, and 146 in the other chain. The next step was to work out the order of the units in the hemoglobin of man, chimpanzee, gorilla, monkey, horse, and other animals. The results made evolutionary history:

MAN AND CHIMPANZEE. Every one of the 287 units in the two chains—141 in one and 146 in the other—is the same. Each unit is in the same place on the same chain, and in the same order. If the 287 units were alphabetical letters forming a sentence—an-

other step-by-step arrangement—the sentences would be the same
in each case, and would, of course, read the same. Such similarity
is beyond the reach of coincidence. It means that man and chimpanzee share the same direct ancestry, and are closely related.

MAN AND GORILLA. In the 287 units there are 2 differences.

MAN AND MONKEY. 12 differences.

MAN AND HORSE. 43 differences.

Other proteins were also tested, particularly the albu-
mins and the transferrins, the latter the iron-binding proteins
of the blood plasma. The results were the same. Man and
chimpanzee had the same or nearly the same lineup of units,
and there was increasing difference between man and the
other animals.

Other highly sophisticated investigations were made.
Vincent Sarich of the University of California, Berkeley, and
others calculated the time required for one difference to occur in a protein. In hemoglobin, they found, one mutation,
or change, tends to appear about every 3 million years. Since
there are two differences in the red blood cells of man and
gorilla, it was estimated that the gorilla had separated from
the stem leading to man about 6 million years ago.

The absence of any difference in the hemoglobin of man
and chimpanzee indicated that the two had been separated
for only a short time. Since the two parted company, enough
time has passed for the evolution of the striking outward
differences in the appearances of the two, but not enough
time for a mutation in the well-protected and little-changing
red blood cells. If man and chimpanzee had not developed
different blood groups, transfusions would be possible.

Until recently it was generally assumed that the line
leading to man had branched off some 25 to 30 million years
ago from some early primate ancestor, and during most of
this century many evolutionists theorized that the ancestor
may have been such a primate as the small tree-climbing
tarsier. However, the new biochemical studies point instead

to a chimpanzeelike predecessor, and indicate that man branched off not more than 10 million years ago and perhaps not much more than 5 million years ago.

About 5 million years ago the forests receded somewhat from their former great expanse. Whether for this reason or some other, some of the primates—chimpanzees—ventured out of the forests onto the drier, more open savannas. Those that could best walk upright—the apes always could take a few steps upright—had their hands freer to hurl a stick or stone at an enemy or prey, as they had always done in a generally aimless way. They were the most likely to survive and leave offspring. Their descendants became the australopithecines, or South African ape-men, and their fossilized bones have been found in southern, eastern, and northeastern Africa in deposits dated from 4 million to 1 million years ago.

To the amazement of scientists who had anticipated that man's immediate ancestors would have large brains and apelike bodies, the fossil remains show that the early forebears were exactly the opposite. Their brains were no larger than those of their chimpanzee predecessors or of living chimpanzees, but their bodies were essentially human, and despite the 400- to 600-cubic-centimeter brains they had learned to chip an edge on a fist-sized piece of stone and to use it as a tool. No ape, not even a bright chimpanzee, had ever mastered this critical skill. Man was on his way.

That way was slow, especially in the early stages. For 3 million years—from about 4 million to 1 million years ago—there was little change in the australopithecines. Their tools, another strong piece of evidence, remained the same simple chipped-edge chunks of stone.

Not until about 500,000 years ago did a human exist whose brain had increased to about 1,000 cubic centimeters, or about twice the size of the brain of the early australopithecines. Java man and Peking man, the men of the larger brain, were unquestionably early men. They chipped somewhat more sophisticated stone tools and knew the use of fire.

An important change had occurred, but again it was only a development of what had gone before.

By 100,000 years ago the brain had increased again. The Neanderthals who lived at that time had heavy jaws and thick skulls. Their brains, though, were in the 1,300- to 1,600-cubic-centimeter range of modern man.

By 35,000 years ago men of substantially the same brain size, but also with the lighter jaws and bones of modern man, lived in hillside caves in the fertile Dordogne valley in France and in other places. At Lascaux and in other caves they painted and incised the walls with some of the greatest assemblages of art that man has yet achieved. Modern man, men like ourselves and named by ourselves—*Homo sapiens sapiens*, or men of wisdom—had arrived.

Since the days of Cro-Magnon man, as the French cave dwellers were designated, man has genetically changed little or not at all. Our brains are triple the size of those of the australopithecines, but in the last 35,000 years there has been no further enlargement or change in the body structure. Man has not become superman.

In the whole 4.6 billion years of our planet, life thus went from the microscopically small carbon chains and globules to the complexities of modern man. That evolution as a process was unbroken, and through it all life was built and rebuilt from the earth's original materials. We are physically made of those elements and no others.

Though the elements were shuffled and reshuffled, used and reused, and up-built a change or a few changes at a time, each form fitted its environment, whatever that might be. We thus became closely and inevitably a part of nature. The two were one, inseparable. Man physically can no more alter this scheme of things and survive than he could take to the breathing of water. For life, or for death, we are a part of nature. More than 4.6 billion years of evolution have made it so.

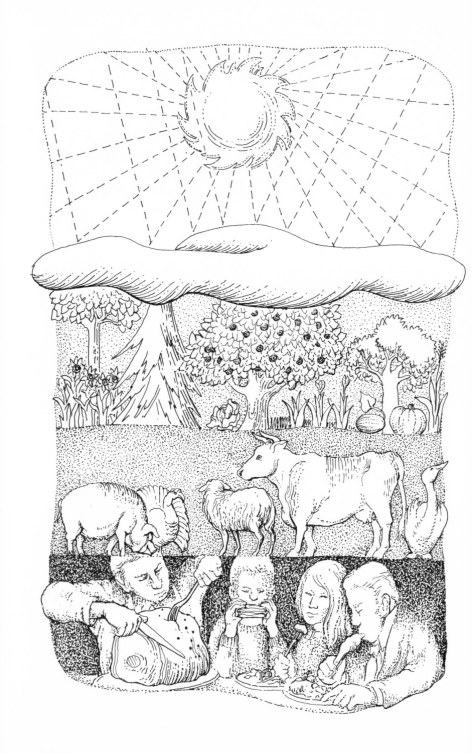

Chapter 3

CYCLES AND DIVERSITY: THE WORKINGS OF THE EARTH

T he earth is neither boundless nor inexhaustible. But for centuries it seemed so to most men.

The first illusion was finally destroyed as *Apollo 8* traveled to the moon and the astronauts—as well as the world watching on television—looked back upon the earth. This was a small planet, as much contained within itself as any spaceship. What men had known intellectually they now knew at last with their own eyes.

In truth, all the elements in the crust or those erupted from the deeper interiors, all the waters in the ocean or on the lands, and all the air enveloping this planet are strictly limited. If these limited elements had not been used and reused, or cycled and recycled endlessly, they would not have supported life for more than a short time. The earth's resources are as finite as the earth itself. They are not inexhaustible.

There is only one significant exception to this law of self-containment and limitation. Only one resource reaches us from outside the confines of this planet—solar radiation. Across some 92.9 million miles the sun pours its energy upon us. It is a resource that never fails, and from the human standpoint, it is nearly without limit.

Abundance, however, is as great a problem for life as scarcity. If there were no shielding by the atmosphere, the uninhibited heat and energy of the sun would probably have prevented the origin of life, and the earth would be a barren, sun-struck globe today. The temperature might well be the 475 degrees measured recently on the surface of Venus.

But the untrammeled heat of the sun does not reach us. As the solar beam passes through the atmosphere it encounters a series of barriers. The first, the high atmosphere, offers a very slight one; only a small amount of the sun's energy is absorbed in this thin air. As the rays next pass through the stratosphere, though, about three percent of the radiation is absorbed.

A greater obstacle is encountered in the lower atmosphere. Still more of the solar radiation is absorbed there, principally by the water vapor in the clouds. Altogether the gases, clouds, and dusts of our atmosphere sequester a little more than 20 percent of the incoming solar radiation.

But in addition to absorbing this fifth of the solar radiation, the atmosphere, and principally the clouds, reflect back another 30 percent. Thus only about half of the sun's beam reaches the surface of the earth and is variously absorbed there. In the polar regions with their great expanses of white ice and snow, for example, most of the sun's radiation is reflected back and relatively little is taken in. In the tropics, on the other hand, where the oceans cover much of the surface, a large portion is absorbed.

There still could be no life as we know it if the earth also did not emit much of the heat it receives from the sun. Some of this absorbed heat is given off by the ground, but even more is emitted by the waters of the earth, and particularly by the oceans, as their heated surfaces evaporate and the rising water vapor condenses in the atmosphere. The condensation releases the latent heat of the vapor, and so some of the heat that once was in the ocean returns to space.

On the whole, these two processes of receiving and emitting heat are balanced. The earth is said to be in a "steady state." If it were not, if over an appreciably long period of time more heat were taken in than is given off, the earth would soon "heat up"—and could become burningly untenable for life. If, on the other hand, more heat were given off than absorbed, the earth would "cool down," and there might be a frozen, icy end for all living things. Only the long-term balance makes life possible.

Though intake and outgo are comfortably balanced on the whole—or have been through all of the earth's history up to the present—there are significant variations. Since the sun beams more directly on the equatorial regions, they receive about twice as much solar radiation as the poles. Thus there is an excess of solar radiation in the tropics and a deficit in the Arctic and the Antarctic. If the earth did not rotate, the effect would be something like heating a fish tank at one end and cooling it at the other. The water would rise at the heated end and flow toward the cooled end, and a violent circulation would be set up.

On the earth some of this potential violence is moderated by the rotation, by the unevenness of the surface, and by the distribution of land and water. But while the circulation is tempered, it still produces the winds and the ocean currents, both of which in turn produce the vertical and latitudinal temperature distribution of the earth. Again, the delicate interrelationship of a number of factors makes the earth the habitable globe it is.

With all of the movements of the weather—the storms and disturbances—the long-term average weather pattern remains relatively stable. Only over still longer periods has climate changed. At times the earth has grown colder and the polar ice has advanced over much of the continents. Again, as the earth warmed, the ice retreated. There have been long dry periods and long wet periods. At times the sea levels have risen and the waters have overflowed much of the land.

Many theories have been advanced to explain the very long-range climatic shifts. Although there can be no simple explanation common to them all, indications are that changes in the earth's orbit, and volcanic eruptions that throw vast quantities of dust into the air, have had an effect. It is relatively certain that only small changes in the mean condition of the atmosphere can produce an ice age, or an expansion of deserts, or a return to other climatic regimes. Everything is linked to everything else. A drop of a few degrees may have caused the last ice age, and it has been estimated that an initial decrease of about 2 degrees centigrade in the mean annual temperatures would again cause an ice sheet to spread from the poles to the equator. Once the ice covered much of the earth, its high reflectivity would result in more reduction in temperature. In such an event, this could become a frozen world. Small changes can produce large results, and the ice will form whether the drop in temperature occurs naturally or should be induced by man's pollution of the atmosphere.

The half of the high atmosphere's solar energy that reaches the earth is the earth's power. By various transformations it becomes the only energy usable by life. Without it no man would move, no deer would bound through the forest, and no plant would grow. Nor could man's machinery work.

The machines, unless they use atomic energy, depend upon fossil fuels, which are entirely the stored energy of the sun, or upon water power, which also depends upon the sun. The earth, man and man's civilization, in short, are solar powered.

Except for a few experimental solar furnaces and solar batteries, man has found that he cannot directly use the energy of the sun. No single human or other animal has ever been born with the capacity to convert the energy of the sun into food; plants alone have this unique facility. In the most literal sense, life depends on the green things of the earth.

From the time the first blue-green algae evolved in the seas, plants have used the sunlight reaching the earth to convert inorganic carbon and water into the complex carbon compounds that are the food and energy of all other life, a process that is known as "photosynthesis." Man has tried for centuries to duplicate it, but has not yet succeeded. The highest skills of modern science cannot accomplish what is done daily by any plant.

In producing the energy that will run the living world, the plants, like any other energy-transforming machine, are not perfectly efficient. They convert only from 1 to 5 percent of the sun's energy they receive into the organic carbohydrates they store in their leaves, stems, and roots or transform into woody material. Even so, the plant biomass of the earth, the total mass of plant substance from lowly algae to great trees, is still enormous—although it has been extensively reduced in the last 10,000 years by the cutting of the primeval forests. Not all of the plant material of the earth is consumed by animals, but its total sets an absolute upper limit to biological production and activity. Beyond that limit, animal life, including man, would inevitably have to halt.

Plants and their biomass are far from equally distributed over the warm parts of the earth's surface. Two large sections have almost none, and produce very little of the total organic energy—the open oceans and the land deserts. The open seas, surprisingly, can be as much of a plant desert as the Sahara. In contrast, some other areas are superproductive, particularly the estuaries, the belts of up-welling ocean along the western margins of continents, the shorelands and fertile islands, the wet tropics, the continental grasslands, forests, and croplands, and the shallow lakes.

Among the waters of the earth, the estuaries with their rich, teeming plankton and attached plants are the biggest producers. But it is these shallow edges where land and water meet that are under the heaviest pressure from man—altogether the nearshore waters and the continental shelves make up only 5 percent of the area of the world. The filling and draining of coastal marshes and pollution are all taking their toll of the energy-trapping water plants.

The wet tropics also produce lavishly. In a year, $2\frac{1}{2}$ acres of jungle may convert the sun's energy into 50 to 80 tons of plant mass. The output of the world's deserts and of the deep ocean, in contrast, is negligible.

The restriction of this green production to certain areas of the earth means that nearly all of the food and chemical energy on which life is based comes from less than 25 percent of the earth's area, and much of it from less than 15 percent. For life, this is a slender, vulnerable part of the globe's surface.

Animals, of course, eat the plants. As they graze on the leaves and the grass, gnaw on the roots and bark, or eat plant debris in the muck and soil, they are taking in the energy of the sun in a form that they can use. With this food or energy they can maintain themselves, grow, and reproduce. Without this food they would starve in a very short time.

The eater may be a cow, an amoeba, or a little whitish aphid, or a deer or a rabbit or a beetle or a man. Man, eating a green salad, is drawing directly upon the converted energy of the sun. The green-eaters, the herbivores as they are called, are legion. There are thousands upon thousands of them, large and small.

Many animals and parasites get their food—their energy —more indirectly. They eat the animals that have eaten the plants; they are the carnivores, or meat-eaters. The lion or the wild dog is typical of this group, as is the lady beetle, a predator of aphids. There is no nibbling on the foliage for them; meat alone is their dish. In some cases they may dine directly on a herbivore, such as a rabbit, but in others they may eat an animal, such as a spider, that has itself eaten a plant-eater. The food energy created by the plants may thus be passed on to three or four other consumers.

Others, most notably including man, eat both plants and herbivores and in a few instances carnivores as well. The energy that man has obtained from the consumers before him has enabled him to spread throughout the world and to build his complex of civilizations.

None of the consumers, however, assimilates or uses all of the food it eats. A large proportion is discarded as gaseous or solid waste, and the animal's own mass is carried into death. At this point a host of new consumers, scavengers, and de-composers move in to eat the energy stored in the dead animal or plant. Fungi, bacteria, earthworms, and some kinds of insects are among them. As they feed on this stored energy they decompose the structure in which it was encompassed. Big pieces are broken into smaller pieces, and in the end the crumbled matter is broken down into the simple components from which it was originally made. Carbon, nitrogen, phosphorus, and the other elements used in living tissue are restored to the environment, ready for reuse by future generations of plants, and through them animals. The material cycle is complete, the elements of the earth having been transferred from plants to herbivores, to carnivores, to decomposers, and finally back to the air, soil, and waters. There may be many steps in the chain, but the elements are up-built, with the use of solar energy, into the most complex of living structures and then broken down again. Without the final breaking down, the earth and all its occupants would be engulfed in waste and the limited store of resources for new life would be exhausted. Only the cycling makes life possible.

The great flow of energy through food chains is not, however, an undiminished stream, all the way through life. At each consuming stage—plant, animal, and decomposer—some of it is lost as heat. Each producer and consumer also uses some of the energy it takes in to maintain itself. Some of this energy is consumed in moving around, in growth, and in reproduction. The great stream dwindles as it goes along, and the effect on life is dramatic.

As the plants convert the energy of the sun into the energy for life, they use only 10 to 20 percent of their production for their own activity. The other 80 to 90 percent produces the huge mass of vegetation.

All of the vegetation lies open to the herbivores, but they do not strip the forests and fields. Although the estimates are bound to vary—cows and leaf-cutting ants have very different ways of grazing—the herbivores probably eat an average of about 10 percent of the world's production of greenery. As leaves fall or plants die, the other 90 percent of the vegetation is decomposed through the work of the detritus food chain and the chemical nutrients made available again to living plants.

The herbivores, like the plants, must use some of their energy for their own growth and maintenance. But in the end only about 10 percent of the food they consume is converted into their own flesh, blood, or other tissues, or into energy available to other consumers. The reduction by this time is drastic. Ten percent of 10 percent is 1 percent; thus only 1 percent of the original plant material has been converted into herbivore flesh. If the original green capital were the equivalent of $10, only a dime's worth would be available for those who eat the herbivores.

The next consumers, the carnivores, are always hungry, and they are efficient hunters. The weasel streaking through the fields may capture and consume about one-third of the meadow mice living there. Other predators also capture the mice. In the end, probably, most mice are eaten, though enough survive long enough to produce another generation, or there would be no field mice.

One conclusion is inescapable: All the energy we receive from the sun and all the essential elements of the crust would not have supported life today nor the great progression of life (1) if energy and elements had not been used and reused through all the 3 billion years of life; (2) if these natural elements had not been used sparingly and with almost no waste; (3) if the great energy, food, and elemental cycles did not work with an almost incredible precision, linking air, water, land, and living things from plants to man to microbes. No one could exist alone. Each is tied into its own wondrous chain, and no breakages are possible or all would be lost. Man has seldom realized what supports him and what he supports, or how completely he is a part of nature.

Although the system has functioned now for some billions of years, although it is worldwide, although it has produced the multitude of living things, man must recognize that it is not indestructible or inevitable. Indeed, in many ways it is fragile, and the margins on which man exists are frequently narrow. We are totally dependent on others—plants—for our food, and all that we have amounts to less than 1 percent of the plants' initial production. Some 5 percent of the world's land, the shores, produces an undue share of our organic energy—and bacteria living in the mud and innumerable other species determine our very existence. It could be touch and go.

A study has shown that the starfish and snails hunting in the shallows along the seashore may eat almost the entire production of barnacles. Predators as a whole eventually capture from 30 to 100 percent of all herbivores, in contrast to the 10 percent of all plants eaten by the herbivores. If these numbers sound illogical, the reason is that when predators consume herbivores, most of the energy they take in is burned in their day-to-day activities, in hunting, moving to shelter, and so forth. It has been estimated that only about 10 percent of their intake is converted into new biomass or into their own flesh, blood, and bone.

Men are typical carnivores. When we eat herbivores, such as cattle, we convert about 10 percent of the cattle flesh into our own bodies. Since the cattle converted only 10 percent of the plant food they consumed into their bodies, we, like the others, have available to us in this case only 1 percent of the original plant food. If we eat carnivorous fish (red snapper, trout, and so on) that have themselves fed on herbivores (small fishes, aquatic insects) that have fed on aquatic plants, we receive even less of the original plant energy (10 percent of 10 percent of 10 percent = 0.1 of 1 percent).

The food supply left to the decomposers is slim if it comes through the food chain. However, much of the unconsumed biomass of the plants falls to them directly. The decomposers do not suffer, but in contrast to the herbivores, whose eating takes so little that it is scarcely noticeable in the forest, the decomposers consume almost all. With several exceptions, no organic compound is overlooked or left uneaten.

The exceptions, however, halt early completion of the majestic cycle. Under special conditions, organic material may accumulate in bogs and lakes for long periods of time, and such accumulations over still longer periods may be compacted into coal, or distilled into oil and gas. These are the fossil fuels that man has been burning at a large rate for the last hundred years. When the fossil fuels are burned, the energy stored in them is at last returned to the atmosphere, in the form of heat. The return, in such cases, is made, but only after a delay of many years, probably millions of years.

Lucretius observed some twenty centuries ago that nothing can be produced from nothing. In the same sense a large supply makes largeness possible, and when material substance is to be spent the only way to spend more is to start with more. (Only in such areas as abstract art and scientific ideas is less ever more.) Because the sun pours a vast amount of energy on the earth, the supply of plants can be large. Frank B. Golley, of the University of Georgia Institute of Ecology, has estimated that all plants of the earth produce about 143.8 billion metric tons of biomass—their own mass—each year.[1] Of this 88.8 billion tons are in terrestrial plants and 55.0 billion tons are in phytoplankton in the oceans.

Since herbivores capture no more than 10 percent of this vegetation and convert only one-tenth of this into their own substance, the production of herbivore biomass amounts to about 1.438 billion tons annually. Inevitably, the mass of cattle and other herbivores is less than that of the plants on which they feed.

The carnivores that live on herbivores manage to produce much less animal substance—or meat. Compared to vegetation eaters, they are relatively few. As a Chinese proverb says, "One hill cannot shelter two tigers." If there were more tigers than deer, the tigers would soon run out of food. The fish and animals that prey on other carnivores and can live only on their flesh must be fewer still. And they are.

What nature has built is an inverted step pyramid, with a large, wide base and sharply terraced sides. On land there are seldom more than three or four steps in the pyramid—the plant, herbivore, carnivore—and in the seas almost never more than five. By the time the solar energy fixed by the plants has passed through three, four, or five successive levels it has been almost completely dissipated, and the decomposers complete the dissipation.

Near the University of Michigan, a field that was once

[1] Frank B. Golley, "Energy Flux in Ecosystems in Ecosystem Agriculture and Function," in *Proceedings of the Thirty-first Annual Biology Colloquarum*, edited by J. A. Winston, 1972, pp. 69–90.

cultivated was permitted in 1920 to "go back to nature." Golley has studied part of the energy pyramid there. The grasses that now cover the field grow lushly in the summer sunshine. About 1,350 mice feed on the grass and plants, and about 130 weasels on the mice, Golley found. The mice must outnumber the weasels about ten to one, if there are to be weasels, for it takes about that many mice to feed one weasel. The numbers of each species are strictly controlled by the amount of food available to them, and since the grass is plentiful, there may be many mice. Where there are fewer mice, there must be fewer weasels.

In such a food chain each successive eater is larger in size. In the seas the mass of phytoplankton is enormous, but the individual organisms are so small they can be seen only under the microscope. On the other hand, the zooplankton, the shrimp and worms feeding on minute plants, are larger, though less numerous, while the small fish that live on the zooplankton are easily visible to the naked eye. Schools of them can be seen swimming through the shallows or around the coral reefs.

The salmon, cod, and the other fish that feed on the small fish are much larger, though again fewer in number. And the large fish at the top of this five-step pyramid weigh hundreds of pounds. The largest of sea creatures far outstrip their prey in size, but significantly the largest animals of all, the whales, eat zooplankton, not fish. Probably they have to bypass some of the intermediate carnivores in the food chain in order to have access to a large enough food supply.

From the billions of microscopic algae and plankton at the bottom of the pyramid to the whales at the top—all are as large and as numerous as the handing down of the sun's energy and the use of the earth's elements permit them to be. Given the kinds of animals and plants the earth has after 3 billion years of life's evolution, only this gradation is possible. It is built into nature, and man is only a part of it—he does not control it.

The same chain of size and numbers would be expected in reverse as the decomposers consume and break down the product of the up-builders. Numerically it can be proved only when parasites parasitize parasites, and they in "turn have smaller fleas and so on infinitum."

The branch that falls to the ground or the squirrel that dies are relatively large units of organic matter. Invertebrates, such as beetles, attack dead animals and fungi spreading their webs on dead wood. The degraded fibers and bones and the little piles of sawdust that result are decomposed into the simple compounds and elements by the still smaller but more numerous bacteria and fungi. "Dust to dust"—the cycle is completed, from a legion to a few to a legion, from the small to the large to the small.

The decrease from many to few and the increase from the small to the large on the "up" side of the food chain is repeated when certain unusable substances are taken in by organisms. These substances are usually those produced by man, like radioactive elements and pesticides. In the 3 billion years of evolution, life had no prior experience with such compounds, and there is no natural machinery to use them as the life materials are used, or to break them down in the body. If there is also no way to eliminate them, all that a living creature can do when it takes in such substances is to store them. And this is usually what happens with many pesticides.

A minnow living in a lake heavily contaminated with pesticides washed down from surrounding farms may swallow ten units of the chemical; it will store almost the same amount. If a large fish eats a hundred of the minnows, it acquires and will store about a thousand units, and should a fishing bird capture a hundred of the contaminated fish, it would take in about a hundred thousand units. The unusable material is thus concentrated; the food chain has "amplified" it.

The case does not have to be hypothetical. To keep down

mosquitoes, a marsh along the South Shore of Long Island was sprayed with DDT every year for twenty years. When George M. Woodwell and his associates made a study of the area, they found that the DDT residues in the upper layer of mud ranged up to thirty-two pounds an acre.[2]

Many insects, fish and animals continue to live in the marsh. Each year they would take in some of the DDT from the annual spraying. When Woodwell ran tests, the plankton in the marsh waters contained 0.04 parts per million of DDT. Small fish which ate large quantities of the plankton contained 1 part per million. A ring-billed gull swooping down over the marsh captured and ate hundreds and perhaps many more of the fish, and from each it received a little more of the DDT. That gull had 75 parts per million in its tissues. From 0.04 parts to 1 to 75—the progression was typical and inevitable. Many of the carnivorous animals in the marsh had concentrated DDT by a factor of more than 1,000 over the organisms at the base of the chain.

DDT is not exceptional. Heavy metals and radioactive elements can also be concentrated as one level of life subsists on another. For example, Perch Lake in Canada is fed by a stream that picks up radioactive wastes from a Canadian liquid disposal area. During a five-year study, about the same amount of strontium 90 entered the lake each year, some of it remaining in the water and some of it settling to the bottom. The aquatic plants of the lake took up the strontium 90 both from the bottom sediments, where the concentration was 180, and from the water, where the concentration was 1. The plants developed a concentration of 280; and beavers that foraged on the plants, a concentration of 1,300.

The plankton in the lake also took in the radioactive materials, and the clams that fed on the plankton reached a concentration figure of 730. The muskrats that ate the clams developed a concentration of 3,500.

The minnows living in the radioactive water had a concentration factor of 950, and the yellow perch, the fish for which the lake was named, consumed innumerable minnows. Perch bones contained 3,000 parts of the radioactive contaminant.

In each of these food chains—plankton to clams to muskrats, or aquatic plants to beavers, or minnows to perch—the concentration of the easily tracked radioactive strontium had gone from 1 in the water or 180 in the mud to 730 to 950 in herbivores and to about 3,000 in the carnivores.

Long-lived strontium 90 settles in the bone, where its radiation can damage mechanisms involved in the manufacture of blood cells, and in man it can produce certain types of cancer. Man, however, does not eat the bones of perch or other animals exposed to the irradiation, and strontium 90 building up through such a fish and animal chain usually does not reach us. Lest man take too much comfort, however, we do ingest strontium 90 with leafy vegetables and to a lesser extent in the milk of cows that have fed on contaminated vegetation.

Other chains, like that of cesium 137, extend directly to man. This radioactive material with a half life of thirty years behaves chemically like an element acceptable to life. It enters cells without difficulty, and is widely distributed throughout the body.

In Alaska cesium 137 fell on the country with the rains. An Atomic Energy Commission study found it in the lichens growing in the forests and on the tundra. Caribou, which live almost exclusively on the lichens during the long northern winters, took it in as they grazed, accumulating about 15 microcuries per gram of tissue. During the Alaskan winters, caribou meat is one of the principal foods of the Eskimos, and Eskimos who live in the contaminated areas now have a concentration twice that of the caribou. And wolves and foxes preying on the caribou have three times the cesium 137 concentration of their prey.

The concentration, like the sizing and numbering of all living things, is inescapable in a world in which one form lives on another. The actual process, however, is rarely seen. To the human visitor to the marsh, the forest, or the city park, all generally appears tranquil.

Charles Darwin once commented in *The Origin of Species* that nothing is easier to admit or more difficult to bear in mind:

> We behold the face of nature bright with gladness, we often see superabundance of food; we do not see or we forget that the birds idly singing around us mostly live on insects or seeds, and are thus constantly destroying life; or we forget how largely these songsters or their eggs or nestlings, are destroyed by birds and beasts of prey. . . .

It is equally difficult to bear in mind the effects of this organization of life, its determination of number and size and of the very form of the living world to which we belong.

Part of this difficulty springs from the endless ramifications involved. One food chain does not usually operate in isolation from all others, indeed, an animal in one chain may also be a member of a second food chain, and any change that affects this animal will affect both food chains. In fact, interaction between food chains is the rule. An upset at any point in one chain is bound to affect others, and often in surprising ways—a fact that is especially true when man tries to modify natural systems.

Charles Darwin studied one case that became a classic. Part of a English heath, where the cattle had freely grazed for as long as man could remember, was finally fenced and the cattle were excluded. A few years later, when Darwin returned for a visit, he saw that Scotch firs were springing up in multitudes in the enclosed area—yet there were none to be seen on the unenclosed sections of the heath where the cattle still grazed, even when Darwin climbed to several higher points where he could look out over hundreds of acres.

Darwin walked down to the open grazed part of the heath to examine it more closely. Between the stems he discovered a host of seedlings and little trees that had been perpetually browsed down by the cattle. In one square yard he counted thirty-two little firs, one of them with twenty-six rings of annual growth.

"No wonder that as soon as the land was enclosed, it became thickly clothed with vigorously growing young firs," said Darwin. "Yet the heath was so extremely barren and so extensive that no one would have imagined that the cattle would have so closely and effectually searched it for food. Here we see that cattle absolutely determine the existence of Scotch fir."

Before the Aswan high dam was built on the Nile, and while the river annually discharged its flood into the Mediterranean, the catch of sardines in the eastern part of the sea totaled 18,000 tons a year. A few years after the dam was built and the nutrient-rich, muddy flood waters were impounded, the catch of sardines dropped to 500 tons. Whether the sardines could not flourish without the nutrients brought to the sea by the river, or whether the changes in salinity that resulted had an adverse effect, is debatable but no longer relevant. It was only certain that the construction of a dam some hundreds of miles away somehow interfered with the life cycles of the fish.

Other cycles were also upset by the dam. The permanent, stable lake that formed behind its high barrier and the irrigation channels carrying its waters to farmlands along the river provided an unusually favorable environment for aquatic snails. The snails had always been in the region, but before the dam was built most of them died each year in the dry season. With plenty of water all the year around, they multiplied—and so did the blood flukes they served as temporary hosts. As the people used the water, the flukes invaded their bodies and caused a serious, debilitating disease called "schistosomiasis," a disease that is now reported to be spreading alarmingly among the people of the Upper Nile region. Nearly all of the 6 million acres under irrigation are infested.

The interlocking chains reach out in wholly unexpected ways. Few anticipated a connection between a dam and a dread disease, or a dam and ocean fisheries, and the few who did were disregarded. Even less was it expected that the Welland Canal in Canada would affect the catch of lake trout in Lake Michigan or lead to a shoreline mess of dead fish. But sea lampreys invaded the upper Great Lakes via the canal and virtually wiped out the lake trout. This meant that there were no lake trout, which are large, active predators, to eat and control the alewives that also came in through the canal, and as a result the alewives multiplied prodigiously. In periodic die-offs their bodies litter the shorelines, and cities along the lakes have had to spend large sums to clear away the odorous mess. Untold connections run through the largely unknown and unmapped food chains of the natural world, and it is almost impossible to predict the consequences of any change at any one point.

Only in communities—the mixed populations in any one place—have some of the principles that govern stability and instability been learned. The more interrelationships there are in a community, and the more diverse the species, the more complex and interrelated are the food chains. And at least in theory, the more complex and interrelated the chains, the more feedback loops there are to lend stability to the system.

In the more than twenty years the "old field" at the University of Michigan has been studied, a rich diversity of grasses and herbs has grown up in the area. The scientists have also counted more than 1,800 species of insects and about 250 species of birds, mammals, and other groups. Diversity greatly increased when the field reverted to nature.

The diversity of the tropical rain forest is even more

overwhelming. The variety of the vegetation is huge, and hundreds of thousands of species of insects have been identified. In contrast, a well-tended corn field has only one species of plant—the corn. If the farmer does his work, all others are eradicated, and the corn occupies all the available land. The field becomes a paradise for flying corn borers.

Endurance in a community varies with its complexity. The tropical rain forest, with its enormous diversity, has endured for millions of years and will continue to do so, if man does not interfere. Trees may spring up and die, but the change is so adjusted that the forest as a whole continues for century after century. The Michigan "old field" is also long-lasting, though it would gradually revert to woodland if the young trees were not nibbled down by deer from adjacent woodlands. The meadow that replaced the one-time cultivated field changes little from year to year. It endures as a meadow, even under intensive but invisible grazing by grasshoppers and other insects.

The corn field, though, would not survive for more than a year unless the farmer replanted it and kept it weeded. The domesticated species is wholly dependent on man, and vulnerable in its singleness. In effect, the corn field is a one-industry town, and if anything goes radically wrong with the one industry, town or corn field is dead. But if disaster strikes one or even a group of species in the tropical rain forest or the meadow, many others are unaffected and the natural economy continues. Diversity insures continuity of the whole, and with time any stricken group may well recover and come back.

It is the same with the insects. In the corn field, since there is only the corn to eat, only corn-eating insects and those that feed on the corn-eaters can live there. A feast is spread before them—an almost unlimited supply of food. The insects accordingly multiply prodigiously, and unless the farmer can stop them, they and not the farmer will harvest the crop.

They will destroy the corn, and in the end their own food supply.

None of the swarming insects of the tropical rain forest has such an unlimited supply of the foliage on which it feeds. All around are hundreds of other plants that the insect cannot consume. The forest is also full of birds and of other insects that will prey on them. Thus the numbers of any one insect are restricted and the rain forest is seldom swept by pests. It again has the safety and continuity of diversity.

The diverse, long-lived community has an inner balance of members. Rarely does any one species flare up in number. Instead, the proportions remain about the same. In the diverse community of the rain forest or the woodland, the herbivores eat enough foliage to sustain themselves, but do not eat enough to wipe out the greenery. This does not represent any innate wisdom, but the fact that their numbers are sufficiently restricted by predators to prevent the whole food supply from being consumed. Such feedback loops keep populations within control.

If the community is a very simple one, there are not enough feedback loops to make such nice adjustments among populations. As a result, population explosions and crashes are more likely. The corn pest, for example, when there are none to stop it, may easily eat itself, as well as the farmer, "out of house and home."

Each species uses certain resources, and in its turn provides resources for other species. Thus the energy harnessed from the sun and the limited elements of the earth are passed along; they are used and reused smoothly by from three to five consumers. It is now certain that this incredibly intricate system works most consistently when the community is diverse and there are many interrelationships. Stability for such communities, and indeed for all life, is thus assured, and has been so assured through nearly all of the 3 billion years of life. Only as man has expanded the plantings of his single crops, his cities, and his other works has the diversity that underlies stability in the natural world been reduced. Locally and sometimes regionally the loss of diversity has turned innocuous insects into pests and has brought ruin to the land and those living on it. Globally the cost has not yet been calculated.

THE "LIFE" ELEMENTS

The living world, or the biosphere, as scientists prefer to call it, is literally of the earth. From plankton to man, all are members of the living pyramid activated by the sun, and all are fashioned from the same range of elements. The elements, of course, are the original ones of the earth, for with a few generally unimportant exceptions none have been created or destroyed since then.

Not all of the hundred and more elements of the periodic table have been drawn into the construction of life. In fact, only six—hydrogen, oxygen, carbon, phosphorus, nitrogen, and sulfur—are used to any considerable extent, while about fourteen others, according to recent estimates, are used in trace amounts and are necessary for human health. Some of the others—particularly the heavy metals gold, mercury, and lead—are poisonous to life. The radioactive elements tend to destroy any living tissue that gets in the way of their disintegrative emanations.

Of the principal six, hydrogen and oxygen are the most heavily used, making up about two-thirds of all living matter, often in the form of water. Their part in the earth and life was considered in the earlier discussion of how the earth got its atmosphere and waters. The next most abundantly used of the elements are carbon and nitrogen. The other two, phosphorus and sulfur, do not bulk large, but are essential; life would be formless and motionless without them.

As large as is the earth's supply of the six, it would not have sufficed for more than a short time if the elements had not been used and reused throughout the entire history of life. The chemical building blocks are generally cycled from rock to soil to water to air, and from one living thing to another; or it can be said that they move through the atmosphere, the lithosphere, the hydrosphere, and the biosphere. None can be permanently retained by any living thing; at death the elements are sooner or later restored to the form in which they were in effect borrowed by the individual, and are then ready for reuse. This is the essential process called "cycling."

All life compounds are carbon compounds. No cell is without its carbon content, and the word "carbon" is nearly interchangeable with the word life. The carbon cycle that supplies life's carbon, like all other cycles, has no beginning or end. By definition it is a continuum. However, one of life's big reserves of carbon is in the air. The plants of the land and the oceans constantly draw upon it, in the form of carbon dioxide. From sunrise to sunset, as the sun shines down on the earth, they use the sun's radiant energy to combine the carbon dioxide and the water in their cells into compounds called "carbohydrates"—the literal food and source of energy for all other life. Production in the green factories reaches its peak about noon, then slowly declines, until it halts at sundown.

Part of the newly produced carbohydrates must be used by the plant itself, for its own metabolism and functioning, and in its use this portion is converted back to water and carbon dioxide. The greater part of the carbohydrates, however, is stored or used in growth, to produce the substance of the leaves, the stems, and the roots. It has been estimated that the forests of the world may contain between 400 and 500 billion tons of such carbon.

As the animals—the plant-eaters—consume some of these carbohydrates, and as they breathe and move, seeking food, shelter, and mates, they in effect "burn" the carbohydrates.

In a complex way they combine oxygen with the carbohydrates, giving off heat, water, and carbon dioxide, which they exhale. Another part of the carbon obtained from the plants is thus returned to its original reservoir.

The use and reuse of carbon fixed by the plants as carbohydrates is, however, not yet complete. As the unconsumed leaves fall or the uneaten grass withers, or as the plankton dies in the sea and drifts downward, the unconsumed carbohydrates reach the ground or the sea bottom, and there—in the soil and in the sediments—the decomposers are always waiting. The microorganisms break down the carbohydrates, and in that process the carbon is converted back into carbon dioxide and returns to its original reservoirs, the atmosphere and the seas, through the respiration of the decomposers.

In certain cases, however, this early decomposition does not occur. As I pointed out in a discussion of the food chains, some of the fallen vegetal matter is not consumed, but accumulates in swamps and shallows, in time to be compacted into coal or distilled into oil or gas—and in this process to lie buried in the ground for millions of years. Now, however, man is altering both this timetable and cycle by the heavy burning of fossil fuels. We burn enough fossil fuel to return between 5 and 6 billion tons of fossil carbon to the atmosphere each year, mainly in the form of carbon dioxide and carbon monoxide. The carbon goes back to the air, via the smokestack, instead of remaining endlessly in the ground. The effect on the air of this large-scale alteration of nature's carbon cycle will be discussed later.

In the seas, the unconsumed plankton that dies and drifts down through the waters in a gentle rain may be eaten by fish and other life living in greater depths. Any plankton that escapes the eaters settles to the bottom and forms the white or gray or tannish sediments that spread across the sea floors. In places these sediments may reach thicknesses of thousands of feet, and in time—again slow-passing geologic times the accumulating material may be compacted into rock, perhaps into dark shale or white limestone. The cliffs of Dover, for example, are almost pure calcium carbonate, the remains of one-time sea creatures. The amount of carbon in the body of a single bit of plankton, say a strand of algae, is minute, but time, numbers, and the cycles of the sea produce from it the ocean's vast stretches of rock. Not until the sea floors are pushed outward by the upwelling of molten materials along mid-ocean ridges are the calcium-carbon sediments carried to areas where they may again become part of the land. The cycle is prolonged.

Carbon is everywhere that life has touched the earth, but while it seems all pervasive, it is not; in meteorites and lavas it is one of the scarcer elements. Life and the interaction between land, sea, and air that life accelerates are necessary to concentrate it.

Whether the carbon cycle is moving as rapidly as a breath or with millennial slowness; whether carbon is cycling through living things, the air, water, or rocks, the cycles have worked with an incredible precision. Everything always has functioned. Carbon always seems to be available at the right time, in the right amount, and in the right form to make this earth possible. How much human interference can be tolerated without upsetting this most basic of cycles no one knows—but it is a question that will have to be taken into account.

Nitrogen is essential in a different way. Nearly 79 percent of the air is nitrogen, so of all the elements it is the most abundant. Nevertheless, this huge supply—a chemically inert gas—is not widely available in the form that man needs to fertilize his crops and sustain his own life. It is naturally put into usable form only by the intervention of a few plants and organisms, or by an occasional flash of lightning.

Usable nitrogen was scarce until man learned how to manufacture it industrially. At that point, when we could make it for ourselves, the natural limitations were largely circumvented, and the natural cycles were partly bypassed or altered. But before the factories were built, man's best suppliers were the blue-green algae and a few dozen other plants—among them the botanically ancient gingkos and cycads, and more commonly the legumes, including the bean and pea—able to remove nitrogen from the air and incorporate it into living tissues by a variety of chemical pathways. The process is called "nitrogen fixation."

Some of the algae grow in rice fields when the rice is developing. Farmers had always seen the algae there, but only in recent years have scientists established that the green film fixes the nitrogen necessary for the production of the protein in rice. In some parts of the Orient the algae are now being cultivated and distributed to fertilize low-yield fields.

When the "fixed" nitrogen enters the soil, it can be taken up by other plants—and by microbes—and used in building their own amino acids and proteins. The animals that eat the plants then can incorporate the nitrogen compounds into their own proteins. It is in this way that man obtains his own essential supply—and without it he would have no properly formed proteins, the substance of his flesh and blood.

At the death of all plants and animals, their nitrogen-enriched proteins are returned to the soil. Bacteria degrade them into ammonia, carbon dioxide, and water; other microorganisms may change part of the ammonia into nitrates and into gases that escape into the atmosphere. Without the bacteria most of the nitrogen would remain in the soil, in a generally unusable state, and in the end would wash down into the oceans. There it would probably settle to the ocean floors and be locked indefinitely into the sediments. Thanks to the bacteria, however, and to other chemical action, much of the nitrogen makes its return to the atmosphere. The nitrogen cycle—a strong and relatively rapid one—is complete and ready to begin again.

Men learned early that if more nitrates could be supplied, their crops would grow better and more land could be cultivated. To achieve this much desired bounty they often grew legumes—but two problems arose. Despite the high nutritive value, many human societies have resisted living on soybeans and other legume products. Then, too, the legumes are "hard on the soil." To mobilize the high energy needed to fix nitrogen, they draw heavily on molybdenum, cobalt, and other minerals in the soil. If the legumes are to thrive, they soon must be supplied with additional minerals—fertilizers, in effect, to grow the fertilizers.

The nearly unlimited demand for fertilizers could not readily be met until industrial production began. The artificial fertilizer plants of the world are now manufacturing about 33 million tons a year, a volume about equal to that produced by nature, and the manufactured fertilizers are used on fields almost everywhere that modern agriculture is practiced.

As the rains come, much of this applied nitrogen is washed down into streams and lakes, and the effect produced is clearly visible. Algae as well as field crops flourish on the nitrates, and their "bloom" or multiplication has filled and fouled waters that were always clear before. The story of what has happened to some lakes will be reported in a later chapter.

More trouble is feared when the now-doubled supply of nitrates eventually reaches the oceans. C. D. Delwiche, professor of geobiology at the University of California at Davis, has found that in the past the oceans denitrified most of the nitrogen compounds that reached their waters. The nitrates were thus changed into a gas that could escape again into the atmosphere. Will the oceans be able to denitrify the doubled load they will soon begin receiving? Will the nitrogen cycle be disrupted at this point? Science is just beginning to study the problem, to seek the answer.

With modern industry's ability to produce nitrates, no shortage need be feared. The problem arises when a large manmade increase in supply disturbs the ancient, elaborately balanced cycles. Overloading, as well as exhaustion of supply, can produce unexpected and often disastrous consequences. Too much, in the case of nitrogen, may be as bad as too little.

Sulfur is another essential for which life is dependent on the earth. Although relatively little of it is required, that little is indispensable, since no protein, including such prime proteins as the hemoglobin of the blood, can be made without it. Sulfur has been called the "stiffening" in protein, but actually it is much, much more. A protein cannot perform its function unless it is three-dimensionally folded in a certain way. Otherwise it would be only a "tired, little lump." It is sulfur that supplies the chemical bonds that whip all proteins into the shapes essential for their myriad functions.

Although sulfur is the tenth most abundant element in the biosphere and the thirteenth in the lithosphere, it has to be recycled sparingly to meet all the requirements for it. Part of its cycle is carried on in the atmosphere. There sulfur occurs mainly in the form of hydrogen sulfide, a form not generally usable by most living things. However, in the atmosphere the hydrogen sulfide interacts with oxygen to form sulfur dioxide. The dioxide falls with the rain and reaches the soil as a sulfate, a form readily utilizable by plants. The plants thus get the sulfur they must have in the form in which they must have it.

As plants and all organic matter decay and fall to the earth, their sulfur is dissolved in the ground water. In this soluble state, it makes its way into the rivers, lakes, and oceans. In time, though, it settles to the bottom. There it might be locked away for ages, if it were not for some obscure, humble bacteria. The sulfur-reducing microorganisms convert the sulfur into hydrogen sulfide gas, which bubbles up to the surface and into the atmosphere. Down in the soft muds of the bottoms of the seas, lakes, and estuaries there is no oxygen to react with the sulfur. Only the bacteria can bring about the chemical reaction that will keep the sulfur cycle in operation.

Life, in a real sense, then, depends on mud and its oxygen-shunning, sulfur-fixing bacteria. Since marshes, lakes, and estuaries, the main homes of these vital bacteria, are being drained or heavily polluted, the world may soon be forced to defend its mud—upon which life itself may depend.

Phosphorus is another element necessary for life. It does not form part of proteins, but no protein can be produced without it. It is part of what is called the "universal fuel" of living matter, ADP and ATP. Only one single atom of phosphorus is necessary for the two compounds that supply energy to our muscles, but without that one atom there would be no movement, and hence no life; not even microbes could exist. A human would not only be motionless, but unable to take a breath or think a thought.

Unlike oxygen, hydrogen, carbon, nitrogen, and sulfur— the other absolutes—phosphorus does not normally enter and form part of the atmosphere. Only if phosphorus from the rocks is swept up by a strong wind or is carried into the air by sea spray does it get into the air. It originates in the rocks, though there is not much of it in most stone. The supplies in the sea are also meager; they are calculated in parts per million. Man formerly found his most abundant and accessible supplies in the guano deposited by birds on some of the desert islands. Millennia went into building the thick deposits.

Only in modern times has industry succeeded in extracting phosphorus from the rocks. The phosphorus present in rocks works its way into the soil as the rocks decay, and there it is taken up by the roots of plants. When the plants are eaten by animals and directly or indirectly by men, the phosphorus is distributed through the biosphere. Man gets his few essential atoms, and at death the supply is ultimately returned to the soil.

Some of the soil phosphorus, however, is leached out by the rains and washed into the water. When it settles to the bottom, it is acted upon by bacteria, which reduce it to its natural state. Like the other elements that reach the sea floors, it is returned to the land only in the slow movements of sea-floor spreading.

Heavy mining of phosphorus rocks and guano in recent years has added unprecedented amounts of the chemical to the soil and water; indeed, some tests indicate that the quantities of phosphorus in these two repositories have doubled. Plankton and pond weeds, both of which are richly fertilized by phosphorus, have increased, too, perhaps in proportion to the phosphorus supplement.

Where this occurs, a lake's own cycles change drastically. The blue-green algae tend to take over, for they can match any phosphorus they obtain with nitrogen from the air, something most other plants cannot do. And as a result of all these special circumstances, undue amounts of phosphorus seem to be going to weeds and algae. It is a serious question whether so scarce an essential can be wasted in this way.

Only such infinitely intricate workings of nature have brought the earth and life to the present state. Though some of the cycles and diversity can be traced in part, and the principles of the grand process are now understood, the whole, in all its ramifications, is still far beyond our present grasp. It is certain only that the connections in nature run far and wide and deep. It is also certain that a break or disturbance at any one point—in a cycle or in diversity—can have effects that no one can anticipate.

The cycles and the evolution of species have gone on for a time nearly incredible to man—from the beginning of life. The systems have also worked with a reliability that man cannot approach in his designs. Because there has never been a large breakdown, we tend to assume that there never will be. And yet the natural systems of which we are a part are not indestructible. Despite their age, extent, and seeming certainty, they are delicately poised, perhaps even fragile. Man has to be concerned about upsets or destruction at any point.

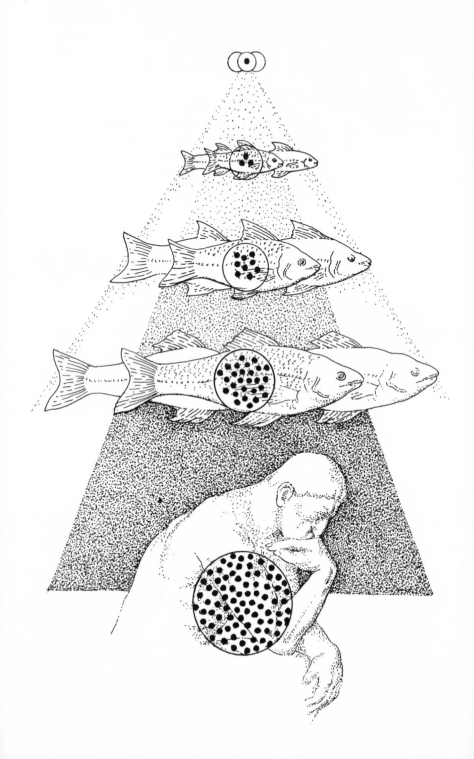

Chapter 4

MAN'S IMPACT ON
THE ENVIRONMENT

The little band of aus-
tralopithecines pursued the weak young ostrich across the
rough savanna. The ape-men were not speedy runners, but
they did run on two legs, and in their hands they carried
sharp-edged chunks of stone, and perhaps clubs; when the
tired young animal was surrounded, a few blows with the
stones must have finished it off. The ostrich was taken back
to the ape-men's camp, a relatively smooth stretch of beach,
before they devoured it, and when the little band—possibly
of twelve to twenty members—moved on, there was nothing
to show that they had been at the lake's edge except the chips
that had fallen when they struck two pieces of stone together
to make their tools, the bones of the ostrich, a few other
bones of young or small animals, and the bones of one of
their own members, who may have died just before they left.
But not even this small disturbance of the site would have
survived if the lake had not risen and covered all of it with
a fine silt that was later sealed in by a volcanic ash fall. Louis

and Mary Leakey found this ancient scene in 1959 as they searched for man's beginnings in Africa's Olduvai Gorge.

It was clear, since nothing was more than temporarily disturbed, that the ape-men had little more impact on their environment than a passing band of apes. During all the 4 million years when the australopithecines spread through much of Africa, into Asia, and across the peninsula that connected the Asian continent with the present island of Java, they merged into nature. Only a few of their bones and chipped stone tools testify to their presence in this sizable portion of the earth. Early man and nature indisputably were one; nature was virtually unchanged by the human newcomer.

Not until about 350,000 years ago did men, whose brains of about 1,100 cubic centimeters had more than doubled the 450-cubic-centimeter brains of their early australopithecine ancestors, learn to use fire. The clay of their hearths has been found in a cave in which they lived near present-day Peking, China. If they then used this new-found tool to burn down the probably menacing forest around them, no evidence has yet been found of it. Not until about 11,000 years ago is there proof of forest fires that may have been set by man. The charred wood and the pollen of fireweeds show that man by this time was trying to alter nature rather than to live with it.

By about 10,000 B.C. settlers in the Middle East's "fertile crescent" and neighboring areas were clearing small plots of land for the building of mud and reed huts. Little clusters of such houses, often with cobbled floors, spread all through the region, into the grassy uplands, and a little later along the Mediterranean coast and even into Greece and Cyprus. The villagers were substituting their own more or less permanent shelter for the caves or temporary shelters of their ancestors. And where they built, the old face of nature vanished.

Since the excavators of the earliest villages have found only the bones of undomesticated animals and wild grains, it is evident that at first the villagers continued hunting the wild sheep, goats, and cattle that abounded in the region and gathering the wild wheat and barley that grew, and still grow, all about.

By 8000 B.C., however, this was changing. Digging into a village of about 7000 B.C. in the Turkish uplands of the Tigris, Halet Çambel and Robert J. Braidwood found grains of a primitive domesticated wheat and the bones of domesticated dogs, pigs, sheep, and probably goats. No longer were animals and cereals taken as they were; man had started to adapt both of them to his own uses, and this was one of the turning points in his relation to nature. Man was discovering that he might control nature—and the discovery was a heady one.

As farms and towns spread into areas more subject to drought, extensive irrigation systems were eventually developed to overcome the dryness. By the fourth millennium the systems were working well, and cities began to supplant the former villages. The Mesopotamians, who were to maintain themselves with stability for the next 3,000 to 4,000 years, permanently changed the character of their land. The earlier vegetation gave way completely to irrigated fields wherever the canals reached. Hydraulic or irrigation civilizations also developed later in North Africa, Peru, North India, China, and Mexico.

Villages also spread up the Danube into lands that were wooded and well watered. At first the settlements were little more than small clearings, but in time the number of these must have been enormous. By 3000 B.C. Europe from the Balkans to Ireland and from Norway to Gibraltar bore the marks of fire and of stone axes.

About 1100 B.C. the iron plowshare, which could cut through thick sod (a characteristic of grassland) as the old stone or wood plows could never do, was developed, making it possible for farmers to move into almost any area. Agriculture spread across the world as acre after acre of soil was brought under the iron plow.

Though the iron plow could break the virgin soil, it did not cut deep enough to cause soil destruction. The new fields remained fertile for some centuries to come, and only gradually did practices creep in that began to destroy the countryside. In Spain the methods of sheep raising turned a land of thick woods and giant trees into an arid wasteland, and when the same methods were introduced into southern Italy, about A.D. 1300, that land was also denuded. With the forest cover gone, both areas were to remain largely barren and unproductive from that time on.

Indeed, the demand for wood was growing everywhere, particularly for the construction of ships. This took a great number of the tallest trees and disrupted many of the remaining forests. Timber was also needed for charcoal burners, for iron-smelting furnaces, for housing, and other city structures. And more farmland was always in demand. Thus, by A.D. 1500, the date often assigned for the beginning of modern times, the forests of Europe were only a remnant of their original stand. Timberlands remained in the mountains and in other relatively inaccessible areas, but generally man had remade the face of Europe.

China and the subcontinent of India and Pakistan were also stripped of their forest during the same period. In India, where the scrub that had replaced the trees was insufficient to feed the goats, the goatherders, with long sticks or billhooks, sometimes knocked or cut the leaves from the remaining trees to feed their hungry animals. More of the trees died, and soon the soil became relatively unproductive.

Where rainfall was low and the sun hot, the cutting of the forest and overconsumption of the grass tended not only to produce barrenness, as in Spain, but desert. Man thus helped to create the Sahara. Earlier the desert had been the home of hunters and herdsmen, whose artists had incised canyon walls with drawings of many of the animals that had once flourished there.

By the twentieth century cities and fields had preempted large parts of the world's temperate zones, where most of the population lives. Man's numbers had grown to 3.5 billion. And by the latter half of the century, the present, man was using, in the sense of managing, about 40 percent of the land surface on the planet. The land vegetation and organic matter had been reduced about one-third.

The inescapable conclusion is that while man has multiplied, his life resources have shrunken.

The estimate that man is using 40 percent of the planet's land surface would indicate that 60 percent still lies open. In actuality most of the 60 percent is too cold, too dry, too mountainous, or too hot for human occupation. It is also generally unsuitable for agriculture. The Food and Agriculture Organization (FAO) of the United Nations has reported after careful investigation that very little of it can be exploited for growing food. Man has already taken over the best land, or about 11 percent, for food crop production, and the best of what remains is only marginal.

After nearly 500,000 years of expansion, or more if the diaspora is dated from the australopithecines, expansion has ended in a large part of the world, and particularly in the developed nations. There, as in the United States, the last frontiers are gone. The FAO predicts that expansion, in undeveloped as well as developed countries, will end by 1985.

With little prospect for further expansion on this relatively small planet—certainly it cannot be enlarged—man must face the additional question: Can he keep the land now in use?

There is every indication that we are literally losing ground. The same conditions that helped to create the Sahara are expanding it southward. Every year two to three more square miles are lost to drought and the sands. The Thar

Desert in India is also spreading, advancing at the rate of one-half mile a year along its whole perimeter. In eighty years an estimated 56,000 square miles, or an area equal to that of Wisconsin, has been turned into shifting sand. Some 2,000 years ago the desert's center was a jungle inhabited by the Indian rhinoceros.

A new semidesert was nearly created in the arid regions of the Southwest in the United States. Despite dryness, the sod had always held the soil in place. But when a period of relatively good rainfall coincided with good prices and a heavy demand for grain, thousands of acres were plowed. Then, in the 1930s, drought came again, and the newly plowed land literally blew away. Dust clouds so enormous that they obscured the sun blew across the Middle West and the eastern United States. In this case the return of more moisture and conservation measures prevented the dust bowl from turning into a wasteland.

Worldwide, the loss of arable land has been steady and telling. Waterlogging and salting are taking other land out of production, and short of drastic corrective measures, destroying it for future use. Such problems in irrigated areas in West Pakistan have resulted in the loss of about 60,000 acres annually.

Fears are also rising that some of the giant new irrigation dams will ultimately reduce rather than increase the acreage available to man. The new high dam at Aswan is holding back the silt that formerly flowed down the Nile and built up the Nile delta. Without the river's silt to restore it annually, some of the delta's past accumulation is being carried away by the sea. The losses, it is feared, may exceed the acres gained by irrigation with the waters impounded behind the dam.

Significant sections of the land that man turned into cropland are also being converted to other uses. For example, the overwhelming growth of metropolitan areas in the twen-

tieth century has wiped out much of the farm and open land
that once surrounded the cities. Potato fields and orchards
are now filled with rows of suburban houses and apartments,
and with shopping center islands in concrete parking fields.

The combined loss of land to drought, salting, manmade
erosion, and urban spread threatens to make serious inroads
into the land available to man. We may in the future have to
do with less land for our livelihood.

With little more land available and with the shrinkage
of present agricultural lands, the world has attempted to
increase food production for the growing population by increasing the yields on the lands now in use. Four methods
principally have been and are being employed: fertilization,
chemical control of insects and weeds, mechanization, and
irrigation. All have an impact on the land.

In the nineteenth century Justus von Liebig worked out
the nutrients required for plant growth. They proved to be,
principally, nitrogen, phosphorus, and potassium, with others
needed in much lesser amounts. If more of these nutrients
were supplied, it was soon learned, plant growth and yield
could be markedly increased.

Manures had always been used for fertilizers, but in the
twentieth century men began to manufacture the chemical
nutrients directly. Even so, heavy, intensive use of synthetic
fertilizers did not start in the United States until the 1940s,
when World War II spurred the demand for more food. Crop
yields doubled, even tripled and quadrupled, and an agricultural miracle seemed to be in the making. This was not lost
on farmers in other parts of the world; they too hurried,
whenever they could, to adopt the new chemical fertilizers,
and by the 1960s the annual world use exceeded 33 million
metric tons, of which 3.3 million tons or about 10 percent
of the total was used in the developing countries. The use of
nitrogen alone increased 543 percent in the twenty-two years
between 1946 and 1968, and world production of the three

principal fertilizers doubled or tripled in each decade. Yields per acre shot up.

Only gradually did the world discover that the dazzling revolution had another, and a costly, side. Fertilizers could destroy as well as increase. With this knowledge, hidden costs emerged.

The fertilizers, as was previously noted in the section on nitrogen, did not stay entirely on the fields to which they were applied; they also washed off into streams and lakes. In the waters they fed algae as stimulatingly as they had corn or wheat, and in their turn, the multiplying algae consumed so much of the water's oxygen that many game fish and other life in the food chains died. Lakes that had been clear and clean for thousands of years began to fill with the slimy green. Lake Erie, which with the other Great Lakes constitute America's largest reservoir of fresh water, was said to have "died."

In fact, only the purer water and the commercial and game fish food chains were being killed. Lake Erie was alive as never before with algae and other organisms that could flourish in the degraded environment. It was not literally "dead"; it was a thriving sewer. Between 25 and 125 feet of muck, most of it loaded with nitrogen and phosphorus, accumulated widely on the once-clean bottoms, and beaches were posted with warnings to stay out of the polluted water. To purify water for drinking took more and more chemical treatment. Other wastes, especially industrial, had contributed part of the pollution, but fertilizers and detergents were responsible for most of the "bloom" of the algae or the eutrophication, as the process is called. All of it together went far to degrade a great lake, as it had been since its waters accumulated in the melting of the glaciers some 11,000 years earlier.

On land, wherever man planted his crops or turned nature to his own uses, he supplanted the complex, intermeshed, many-specied life of the forests and grasslands with a few species of his own choosing, particularly with corn, wheat, and cotton. The relatively few species of insects that could eat the selected crops had a rare opportunity. Acres upon acres of their favorite food stretched out before them, and with a nearly unlimited food supply, they multiplied prodigiously—and became pests, as species in the wild rarely did. Men sometimes also inadvertently imported the pests from other places or continents, with the result that, on a new continent with a big food supply and without their traditional enemies to hold them in check, some of the newcomer pests multiplied even faster than the native insects. Farmers' losses from such insects as the boll weevil and corn borer ran into the millions. Essential crops were often threatened— and so was the human food supply.

All seemed to change in 1948, when Paul Herman Muller of Switzerland proved the effectiveness of a compound called DDT, which almost magically killed insects. For what seemed a great boon to mankind and was, in its reduction of mosquitoes and malaria, Muller was awarded the Nobel Prize.

Only gradually did the reports on DDT begin to come in:

In one laboratory shrimp were exposed to DDT concentrations of less than 0.2 parts per billion. In less than twenty days every shrimp was dead. The laboratory reported that concentrations of this magnitude had been detected in Texas rivers flowing into commercially important shrimp nursery areas.

In California the production of Dungeness crab fell as DDT residues were observed in the developing larvae.

The bald eagle and the peregrine falcon disappeared from the wilderness of the Channel Island off the coast of California. In the same area the brown pelicans and double-crested cormorants no longer reproduced, the eggs in each case breaking during incubation. Only one brown pelican hatched in 1,200 nesting attempts. The birds had fed on fish in which DDT compounds exceeded 10 parts per million, and the DDT had caused a thinning of the egg shells.

Mackerel caught off the California coast exceeded the DDT tolerance levels permitted for human consumption.

Reproduction of freshwater game fish in Lake Michigan was threatened.

The Study of Critical Environmental Problems (SCEP), a conference and work force of experts sponsored by the Massachusetts Institute of Technology, estimated that about one-fourth of the 63,400 metric tons of DDT that the United States produced in 1968 finally made its way to the oceans, either through runoff or through transportation in the air.

Part of the tremendous input of the pesticide is quickly ingested by the little fish, which are eaten by the bigger fish, which may be eaten by still larger fish, and by man. If the second eater consumes many of the little fish, as demonstrated in the food chains, these fish take in multiple supplies [1] of DDT, and the larger fish, eating large numbers of the middle-sized fish, consume still greater multiples of DDT. The linked food chains make the outcome inevitable. Great concentrations of DDT accumulate in the fish and other animals at the end of food chains.

Actual tests showed that gray and sperm whales, at the end of one of the chains, contained from 0.4 and 0.6 parts per million in their blubber—DDT accumulates in fat. Sea birds, at the end of a marine chain, had concentrated 10 parts per million and migratory tuna, at the end of still another, up to 2 parts per million. Oysters living along the coast and thus closer to application areas sometimes showed up to 5.4 parts per million. DDT was even found far from its point of application, in the seals and penguins in the Antarctic—the chain of eaters and the atmospheric circulation had extended that far.

Said the SCEP report:

[1] See Chapter 3, pp. 37–39.

From these few observations, it is possible to deduce that DDT and its residues are most probably distributed throughout the marine biosphere. . . . One quarter of the world's production of DDT may have entered the oceans. . . . The oceans are the ultimate accumulation site. . . . DDT has already produced a demonstrable impact upon the marine environment.

If the accumulation should continue, as it must as long as fish eats fish and fish eats plankton, the scientists warned that there might be "no opportunity to redress the consequences." They also cautioned that their predictions of the hazards might have been greatly underestimated. They were saying that the teeming life of the oceans and the chains of life built up in more than 3 billion years of evolution could be destroyed or irreparably damaged in a few years. If, because of their greater accumulations of DDT, the large predators at the ends of the food chains were to disappear, the regulation of populations of smaller fish would be disrupted, with unpredictable results. The smaller fish might increase and exhaust their food supplies, with resultant population crashes —the boom and bust regime.

The evidence also began to show that DDT and other pesticides were increasing the number of pests, and in the inevitable chain reaction, altering all life in the areas treated— including that of man. For example, in the coastal Cañete valley of Peru, the cotton fields were first sprayed in 1949 with DDT and other chlorinated hydrocarbons. Soon the cotton yield increased from 440 pounds per acre to 648. Elated, the farmers applied the insecticides "like a blanket." They even cut the few trees so the spraying planes could maneuver more easily.

Three years later the DDT was no longer effective against aphids. The first cotton pests had largely been killed, but resistant strains—mutant strains unaffected by DDT and the other materials—had arisen. The mutants had few natural enemies to hold them down, and they multiplied. A few years later a boll weevil infestation reached a new high, and six new pests had appeared as well. The latter were not pests in the nearby unsprayed valleys, where natural enemies kept their numbers low. In the sprayed areas the natural enemies, as predators eating many of the insects, could not escape—they were killed by the DDT they accumulated. Five years after the spraying started the cotton yield was about one-third lower than it had been before the spraying started. Economic disaster came to the valley.

California cotton growers spent thousands of dollars to rid their fields of the lygus bug. After the work was all done, they found that the reduction in numbers of lygus did not lead to increased yields, seemingly because the bug fed largely on "surplus" bolls of cotton that would not have ripened in any case. But the spraying also killed predators of the boll worm. The worm had escaped the spray, and without its predators it began to increase. Here the spraying brought back an old pest.

Similar disasters occurred throughout the world, and scientists called in to try to remedy the havoc always found that the number of species had been reduced and the ecosystems simplified. The runaways, the outbursts of the new pests, were entirely predictable. With the predators removed either because they were killed directly by the sprays or were poisoned by eating contaminated insects, the way was open to the survivors. Outbreaks were certain to come. When many species are present, attack and defense are balanced.

Agriculture nearly always reduced the complex network of nature to a few species, and insecticides reduced it further. Resistant species and those suddenly relieved of their enemies took over in many areas.

Agricultural surveys confirmed that in the long run the pesticides did not reduce the amount of the crops lost to insects. Different insects then preyed on the crops, and their numbers changed, but insects still consumed about 10 percent of the crops man grows for himself. Broadcast insecticides brought no gain in the general battle against pests.

As part of the continuing struggle to produce more food, man also attempted genetically to develop crops with higher yields. One new rice matures in 120 days as against 150 to 180 days for the old strains. Hybrid corn multiplies former yields. Cattle whose early ancestors once gave 600 pounds of milk a year now on the average produce 9,000 pounds annually. The "green revolution," as the rise in productivity is called, is sweeping the world, in the undeveloped as well as in the highly developed countries. It initially brought new prosperity to Mexico and others it reached.

Economic studies are beginning to show, however, that the extra energy added in fertilizers and the energy—fossil fuels—required to operate cultivating and harvesting machinery just about equal the extra energy obtained in the higher yields. Without heavy subsidies of energy, the higher yields cannot be maintained.

If agriculture has expanded about as far as it can go, and the use of fertilizers, pesticides, and the new genetics of the green revolution are not solving the problem of more—more food, more energy, more of everything—many insist that man himself will make what he needs. He has already gone far in this direction.

Synthetic fabrics have partly replaced cotton and wool; detergents have nearly driven out soap; many cars roll on synthetic rather than natural rubber; and plastics have partly routed wood and paper. The implication is that modern civilization can very well do without natural products if it has to.

Barry Commoner, ecologist of Washington University in St. Louis, has called the concept into question and linked it to the impact modern manufacturing is having on the environment. Both cotton and nylon consist of long chains, made by linking together small units, or monomers. To bring the links together in the chain takes energy. The cotton plant obtains the energy it uses in the assembly from the sun, and makes the transformation at low temperatures. There is no cost, and there is no pollution.

Commoner said:

Compare this with the method for producing a synthetic fiber. The raw material is usually petroleum or gas. Both represent stored forms of photosynthesis energy, just as does cellulose. However, unlike [the cotton plant's] cellulose, these are nonrenewable resources in that they were produced during the early history of the earth in a never-to-be repeated period of very heavy plant growth.

Moreover to obtain the desired monomers from the mixture present in petroleum a series of high temperature, energy-requiring processes, such as distillation and evaporation, must be used.

All of this means that the production of synthetic fiber consumes more non-renewable energy than the production of a natural fiber such as cotton or wool. It also means that the energy-requiring processes involved in synthetic fiber production take place at high temperatures which inevitably result in air pollution.[2]

All of the synthetics thus cost more in terms of energy consumption, Commoner maintains. No factory can match the economy of a natural plant. Nor does any factory directly use the nearly inexhaustible and free energy of the sun. The supplies of oil, gas, and even coal are both limited and potentially exhaustible, and certainly reduced by the manufacture of synthetics.

The synthetics create other problems, particularly on the waste heap. Plastics may last for years. They are new to life, and few, if any, microorganisms exist with the capacity to degrade them. Plastics have become the most common type of flotsam on the world's oceans, and quasi-stable concentrations of as many as 3,500 pieces per square kilometer float about in the Sargasso Sea.

[2] *Environment*, Vol. 13, No. 3, p. 13.

Making it in the factory or with industrialism does not, then, produce a certain answer to meeting many of man's needs. Indeed, every new way taken so far to provide more foods and other products for the billions now living on the earth has run into difficulties. And the prospect for success is not promising, as a review demonstrates:

1. The land of the globe is not expandable.

2. Most of the arable land is already in use. Little more can be brought into production.

3. The present manner of using fertilizers to increase production threatens to destroy much aquatic life and to pollute the waters.

4. Known pesticides fail in the long run to hold insects in check and may destroy the ecological systems of which man is a part.

5. The "green revolution" takes as much in energy as it produces.

6. Synthetics cost more than the natural products of the soil and contribute to the pollution that is choking land and air.

From the time of the first reed-and-mud hut villages in the valley of the Tigris and Euphrates, cities and industry have undeniably changed the environment. In this century the change accelerated, and in some industrialized cities has reached a critical point. The effects on land, air, and water have also become global as well as local.

An estimated 3,000 chemical components have been added to the atmosphere, and perhaps 500,000 polluting substances have reached the oceans. The trash cast off by cities has reached mountainous proportions; recent surveys show that waste materials cover about 210 square kilometers of the bottoms of the New York harbor and the adjacent continental shelf.

The SCEP conference took particular note of the carbon dioxide the burning of fossil fuels is pouring into the air, but was uncertain whether this or other pollutants accounted for the rise in global temperature from 1901 to 1945 and for the present cooling trend. Said the SCEP in its report "Man's Impact on the Global Environment":

Man can become so numerous and powerful that his activities will change the composition of the atmosphere or the character of the earth's surface enough to produce effects larger than those "naturally" expected. Ice advances and retreats, wide-spread droughts, changes in the ocean level and so forth, were accomplished by only slight shifts in the mean circulation patterns and only small changes in the average temperature over large parts of the earth. . . . A change in a mean condition of the atmosphere may be accomplished by a relatively small perturbation . . .

Are there leverage points? Could some relatively small manmade change significantly alter global climate?

The SCEP scientists concluded that the carbon dioxide in the atmosphere is increasing now about 0.2 percent a year, and that about half of this load remains in the atmosphere. The other half is taken up by the biosphere and the oceans. Whether the carbon dioxide will increase to the point where it would absorb enough solar radiation to melt the ice sheets and flood the world's seaports, no one knows. The SCEP cited a potential danger and called for "continuous analysis."

Another danger point lies in fine particles carried into the air from such natural sources as volcanoes, dry lands, and plowed fields, and increasingly from smokestacks and exhaust pipes. According to their color, the fine particles either reflect or absorb radiation from the sun and thus affect the heat balance of the earth. In the first ten miles of the atmosphere, still finer particles may be the nuclei around which droplets collect; thus they become the mechanisms by which clouds, rain, and snow are formed.

Measurements at twenty nonurban sites showed a 12 percent increase in particle concentration in the air between 1962 and 1966. Will the particulates block enough of the sun's radiation to turn the earth into a gloomy planet where sunny days are a rarity? The SCEP pointed out that any modification in climate produced by the buildup in particles —manmade particles are now about one in five—will first become manifest locally, well before any global change is felt; the establishment of monitoring stations was strongly recommended. The SCEP scientists predicted that man should have the time to reverse any adverse trend by promptly taking control measures.

Supersonic and other aircraft flying in the troposphere— the first ten miles of the atmosphere—have already increased cirrus cloudiness, particularly in the eastern part of the United States. The water vapor and particles they emit could also affect the ozone that shields the earth and could contribute to further cloudiness. The main danger here is that the jets' hydrocarbon emissions might react with sunlight to produce smog. It is known that after the 1963 volcanic eruption of Mount Agung in Bali and the injection of a large amount of dust into the atmosphere, the temperature of the lower equatorial stratosphere—the upper reaches of the atmosphere—rose six or seven degrees, probably because of the absorption of solar radiation by the particles. Could the emissions of planes produce a similar effect? The scientists expressed a feeling of "genuine" concern about SST flight.

Still another trigger that could set off large changes in the atmosphere and hydrosphere is the spilling of oil. Studies indicate that about 2 million metric tons of oil are introduced directly into the world's waters each year. About 90 percent comes from the operation of tankers and submarine oil wells, as well as from the disposal of spent lubricants and the fallout of hydrocarbons emitted by vehicles and industry. It is predicted that, with the larger tankers now planned and increased drilling on continental shelves, about 4 million tons of oil may enter the oceans annually by 1980.

The oil has a more drastic effect on the life of the sea than upon climate, but it might possibly interfere with evaporation and rainfall. It directly poisons such filter feeders as clams, oysters, fish, and many birds, and affects all the life dependent or feeding on them.

When the oil tanker *Torrey Canyon* foundered off the southwest coast of England and when the leak developed in a Santa Barbara, California, offshore oil well, additional damage was caused by the detergents used to disperse the oil slicks. The slicks picked up DDT and other hydrocarbons already in the water, for these chemicals are soluble in oil. The floating organisms that comprise the plankton of the sea could not avoid the DDT-soaked oil slicks, and they were either killed or the DDT was absorbed and concentrated as it moved up the food chain.

While the global effects of pollution are only now becoming clearly perceptible and sometimes acute, the pollution that overhangs the cities is all too evident. A traveler approaching Chicago by air on some days will see a pinkish gray umbrella of smoke hanging over the city. Nearing Nagoya, Japan's famed bullet train runs into what seems to be an inverted bowl of black air. The city is embedded in its polluted air, like a specimen in plastic.

But virtually every metropolis has an air pollution problem. Pollution now cuts down the sunlight reaching New York City. On the other hand, when London reduced its pollution, by reducing the consumption of smoky fuels, its sunny days increased.

According to the United States Public Health Service, the 90 million motor vehicles of this country annually pour out 60 million tons of carbon monoxide, 1 million tons of sulfur oxides, 6 million tons of nitrogen oxides, 12 million tons of particulate matters, plus lesser amounts of other dangerous substances. In addition, industrial plants emit 2 million tons of carbon monoxide, 9 million tons of sulfur oxides, 3 million tons of nitrogen oxides, about 1 million tons of hydrocarbons, and 3 million tons of particles.

The fuel burned for heating homes, apartments, and offices adds another 2 million tons of carbon monoxide, 3 million tons of sulfur oxides, 1 million tons of hydrocarbons, and 1 million tons of particulates. Trash burning contributes still more particles.

All together, cars, factories, and homes put 140 million tons of pollutants into the atmosphere each year, or about three-fourths of a ton for every man, woman, and child. How much can the atmosphere take?

Man's impact on his planet is overbearing. A question raised by UNESCO—will the impact make this earth uninhabitable?—must be faced. It can no longer be ignored.

Chapter 5

MAN'S IMPACT
ON MAN

About 300,000 years ago, as our early ancestors were beginning to develop considerable skill in chipping stone tools, the population of the earth may have reached a new high—about 1 million. Now some 3,000 centuries later the number of humans living in nearly all of the habitable parts of the earth has soared to more than 3.5 billion.

From 1 million, if that is taken as a starting point, to 3.5 billion on a globe that is exactly the same size is both phenomenal and troubling. And if the increase continues, it could ultimately be lethal.

The increase is even more phenomenal if it is studied closely. Most of it, in fact nearly two-thirds of it, has occurred in the last 100 years. A table shows how it came about:

300,000 years ago	1,000,000
10,000 years ago	5,000,000
100 years ago (1870)	1,000,000,000
(1970)	3,000,000,000+
The present	3,500,000,000+

However startling such an increase may seem to man, who sees the crowding around him and tries to comprehend what it means to have 3.5 billion humans on this earth, neither the numbers nor the acceleration are unusual in themselves or prodigious in the scale of nature. Rather, this is a common pattern of nature.

In 1798 Thomas Robert Malthus, in his *Essay on the Principle of Population*, was one of the first to demonstrate that "the constant tendency in all animated [life] is to increase." Malthus argued to the still unabated anger and dismay of many that there would be no limit to that increase unless, in effect, it were checked by hunger, disease, war, or some other action. Without some check the increase would be geometric.

Charles Darwin emphasized the same pattern. The tendency to geometric increase is universal. Unless checks intervene, no district, no station, not even the entire surface of the earth could accommodate even the progeny of a single pair after a certain number of generations. In his day even the numbers of slow-breeding man had doubled. Darwin himself counted the seeds of an English orchid. By laying them on a long ruled line he found there were 24,080. If nothing had interfered, a fourth generation would have covered the whole surface of the earth with a uniform carpet of green.

"Lighten any check, mitigate the destruction ever so little and the number of the species will almost instantaneously increase to any amount," Darwin wrote in *The Origin of Species*.

Richard S. Miller of Yale University was one who theoretically applied the rule to bacteria. A single bacterium splits in two in twenty minutes. If there were no constraints and no limits in the environment, a colony of the bacteria one foot deep would stretch across the face of the earth in a day and a half. An hour later the colony would be over the heads of hapless humans. In a few thousand years it would weigh as much as the visible universe and would be expanding with the speed of light.

But neither bacteria nor orchids nor man have yet smothered the earth. All have, so far, stopped short of such engulfment. Indeed, the numbers of all, except man, have, despite occasional runaways or spurts, remained relatively constant. No one species has yet inherited the earth, and the balances have generally held up to the present—except for man.

At first man's numbers were few, limited to a few bands wandering the savannas. Deaths must have nearly equaled births, and it probably took some 1 million years for the population to reach an estimated 5 million.

Later still agriculture developed. No longer were some ten square miles of hunting territory needed to provide a family with meat and other sustenance; a few acres of planting would do. One major check on the increase of population, the scarcity of food, was lightened, and the population "almost instantaneously" increased. According to some studies it multiplied by sixteen in about 4,000 years following the spread of agriculture.

Disease and other hazards of life still were checks on geometric increase, as at least half and perhaps two-thirds of the children born died before they were old enough to produce offspring of their own. Between 1348 and 1350 the bubonic plague killed about 25 percent of the inhabitants of Europe, and at about the same time, 1348 to 1379, the black death cut England's population from 3.8 million to 2.1 million.

Gradually, though, life grew a little better and less hazardous. Farmers discovered that fields could be renewed by planting red clover—it restored lost nitrogen—and more food was produced. The plague at last subsided and there was relative peace. From 1650 to 1750 it was estimated that the population of Europe and Russia increased from 103 million to 144 million. In Asia, too, there was a little more stability, and the population there also apparently soared.

In the next century, the 1800s, the industrial revolution, sanitation, and vaccination reduced the death rate, and the population of Europe doubled. In contrast, that of Asia, where there was little industrialization, increased only 50 percent.

"Mitigate the destruction ever so little," Darwin had said in 1859. And man was proving him right. A few years later, the world population reached more than 1 billion.

Increase came even more rapidly in this century. Food production increased. Infectious diseases, particularly yellow fever, malaria, smallpox, and cholera, were brought under control. At the same time the practice of medicine improved. Millions of babies, who a few centuries earlier would have died in infancy, survived to produce their own children. Death rates had previously fallen somewhat, but now they declined drastically, a drop primarily reflecting the survival of more of the young rather than any great decrease in the death of the old, or their survival into an eighth or ninth decade of life.

The story was the same in many tropical and undeveloped countries. In Ceylon, where malaria had always killed many of the young and some of the old, spraying with DDT was started in 1945. The death rate, which had been 22 per thousand, by 1969 stood at 8 per thousand, and the population of the island multiplied—for population is determined not only by how many are born but also by how many die.

Many of the twentieth century's surviving children grew up and flocked to the cities of the world. In the highly industrialized cities of the West, some of the urban newcomers found that large numbers of children were not economic assets, and the birthrates declined. In Denmark, Norway, and Sweden, where the birthrate had been about 32 per thousand in 1850, it dropped to 16 a century later. Similar reductions occurred in

England and France, but not in the nonindustrialized states.

In the United States, though fewer were being born—except during the "baby boom" that followed World War II—many more were surviving to become parents. Suddenly there were more in the fifteen- to forty-four-year age bracket —the childbearing years—and again there were more children. The large proportion of young parents in the postwar populations thus sent populations shooting up again.

Such changes were most clearly visible on the tight confines of an island—though the earth itself is an island in space. On Mauritius, the Indian Ocean island long known as the home of the extinct dodo, 44 percent of the 800,000 population is under fifteen. The island is struggling to provide schooling and care for the thousands of children.

Another factor entered into the present population increase—the sheer size of modern populations. When 2 become 4 the impact may scarcely be noticed, but when 1 billion become 2 billion and the 2 billion turn into 4 billion (the prediction for 1975), cities become jammed, school crises develop, pollution mounts, and the earth is in shock. When populations double, food, goods, and services must at least double to maintain existing standards of living.

Demographers bring the population factors together and obtain a meaningful measurement by computing the "doubling time," or the time it takes a population to double in size.

In the early years of man's history, when the population of the earth was very small, it doubled every 1,500 years. The Ehrlichs, in their book *Population, Resources, Environment*,[1] demonstrated how the 5 million of 8000 B.C. became 640 million and more by A.D. 1650; in those 9,000 years the population of the world doubled six or seven times. The next

[1] Anne H. and Paul R. Ehrlich, *Population, Resources, Environment: Issues in Human Ecology,* 2nd edn. (San Francisco: W. H. Freeman, Publishers, 1972).

doubling, from 500 million to 1 billion, was accomplished in 80 years.

The doubling in which the world is now engaged, from 2 billion to 4 billion, is expected to take 45 years. If the 1970 growth continues, the next doubling will come in about 35 years. By the end of this century the population is expected to be about 6.5 billion, or roughly twice that of 1970.

These are the people of Spaceship Earth. The overall figures, though, conceal some critical differences among regions and nations.

The population of the United States now is doubling "only" about every 63 years and that of Europe about every 100 years, as compared to a doubling of 31 years for Africa, 32 for Asia, and 24 for South America. It was on the latter continents that the dramatic reduction in the death of infants and children suddenly produced huge new classes of the young. Nor was there much of the urban slowing down that occurred in the more industrialized regions.

Latin America faces a particularly urgent problem. Poverty and malnutrition are widespread, even now when the population stands at 275 million. The possibility of increasing food production in the tropics is not promising—the soils cannot produce the proteins needed.

Asia's population is already about 2 billion. With a doubling every thirty-two years and the certainty, if the present rate of increase is maintained, of 4 billion by the end of the century, food and services will have to be at least doubled to avert disaster.

The one exception to the hectic upsurge of people in Asia is in Japan. The islands of Japan began early to feel the pressure of overpopulation, and in 1948 abortion was legalized—with marked results. In four years the birthrate began to fall, and it has declined from the former 38 per thousand to about 17.

The tremendous increase in the population of Asia and

Latin America—and also of Africa—is changing the distribution of the world population and worsening the disparity of resources.

In 1900, when the world population stood at 1.6 billion, one-third of the people lived in Europe and North America, and the other two-thirds in Asia, Africa, and Latin America. Now, with more than 3.5 billion in the world, only one-fourth live in the industrialized world of Europe and North America, while three-fourths struggle for existence in other areas. If populations should continue to increase as they are doing, by the year 2000 the ratio would be one-fifth to four-fifths.

Miller, in an article, "Man and His Environment: The Ecological Limits of Optimism," pointed to the implications: "Eighty-five per cent of the world population increase of 3,300,000,000 between 1965 and 2000 is expected to occur in Africa, Asia, and Latin America, where malnutrition and starvation already are serious and where severe food shortages and widespread famine are predicted for the future." [2]

The scientist used an analogy to bring home the point. If all the people of the world could be reduced proportionately to a town of 1,000 people, there would be 60 Americans and 940 others. The 60 Americans would have half the income of the town with the other 940 dividing the other half. White people would total 303, and the nonwhites 697.

To continue the analogy, the 60 Americans would have fifteen times as many possessions as the rest of the town. They would produce 60 percent of the town's food supply and either consume or store most of it. Since most of the 940 non-Americans in the town would be hungry most of the time, it would create ill feeling toward the 60 Americans, who would appear to the majority of townspeople to be enormously rich and fed to the point of sheer disbelief. The Americans would also have a disproportionate amount of

[2] Richard S. Miller, Yale University School of Forestry *Bulletin* (1970), No. 76, pp. 14–15.

the electric power, fuel, steel, and general equipment. Of the 940 non-Americans, 200 would suffer from malaria, cholera, typhus, and malnutrition. Few of the 60 Americans would get these diseases, or probably have to worry about that possibility.

The actual figures only underscore the analogy. The United States, with its population of 200 million, has only one-sixteenth of the world population, but it consumes over half of the world's annual production of nonrenewable resources.

"We must realize," said Miller, "that the quality of human life, and in fact man's destiny on this earth, is directly related to the basic equations of population growth in a limited environment." [3]

The predictions cited have pertained only to the next 30 years, or only to the time in which a child born today will reach maturity and become established in a career and home of his own, but some of those studying the figures and trends carry them much further. The world's present rate of population increase, doubling every 35 years, is obviously impossible, for in 1,000 years, or roughly the time that has passed from the first Crusades to the present, we would have 1,700 persons for every square yard of the earth's surface, land and sea. There would not even be standing room.

Up to now all species runaways have been halted short of such cataclysms. Food or territory were eventually exhausted and expansion stopped. Until fairly recently man was subject to disease, starvation, and many of the same checks as other species. Only in the last few centuries has infectious disease been brought under substantial control, and only in this time has production been stepped up in the West until it could nearly supply the needs of all. Our transportation has also conquered geography and the isolation that permitted some disasters to run their course. Thus the historical local checks no longer operate as they did in the

[3] Ibid.

past, and having lightened his load, man is increasing.

Will man intervene again with his own unique intelligence to hold his numbers in check? If not, will new plagues, nuclear wars, the destruction of the environment, or some unforeseen development call the halt? It will have to be one way or the other, for the earth has only 57.4 million square miles of land. The globe can carry only so many—and exactly how many, no one can say definitely. Studies of the optimum population for the world are only beginning, and much may depend upon the impact the billions have on the billions.

If man should begin to feel impossibly overcrowded, as some do even today, unexpected limitations could come into play. Many experiments have been carried out with rats. When rats were crowded in their cages, many of the females failed to give birth—and the young that were born were carelessly nursed, deserted, and sometimes eaten.

TECHNOLOGY AND POLLUTION

At what point man might react against the stress of crowding is not now predictable. Edward T. Hall of Northwestern University has shown that much depends upon culture. Some cultures not only tolerate but seek closeness. In others, particularly the northern European, individuals do not willingly accept any invasion of their personal "space bubble" or the immediate space around them. Hall holds that some personal space is a necessity of life.

However severe the impact of man's numbers on men and the planet, a second great impact comes from technology. Barry Commoner attributes much of the pollution and deterioration—our impact on one another and on the earth—to the technologies developed by affluent societies, rather than to population increase in itself.

In the United States the gross national product—the

sum of personal and governmental expenditures for goods and services—increased 126 percent from 1946 to 1968. During the same time the population rose 43 percent. The amount of lead in the air increased 400 percent, and the nitrate and phosphate in Lake Erie by 220 percent. Smog over large cities multiplied tenfold. Pollution, on the face of the figures, was growing even more rapidly than people.

"It would appear," said Commoner, "that the rise in overall United States population is insufficient in itself to explain the large increase in overall pollution levels since 1946." [4]

Closer analysis has indicated to Commoner that a shift in technology, a sudden shift to synthetic rather than natural products, created much of the trouble. It was the new industries and the new products, and their nature, that caused much of the visible pollution and the acute drain on resources.

The production of synthetic fibers, according to Commoner, increased 1,792 percent in the post World War II period, while the use of fibers of all types increased only 6 percent. To produce the new, mostly manmade fibers—nylon, rayon, and other synthetics—consumed far more nonrenewable energy—petroleum, coal, and natural gas—than to grow and mill cotton, wool, hemp, and other natural fibers. The factories also caused more pollution.

It was the same with the synthetic detergents that partly replaced soap, with the synthetic rubber that tended to drive out the natural product, with the plastics that took the place of many wood and paper products, and with the pesticides that largely supplanted the natural ecological processes for controlling pests and unwanted weeds. The shift from natural to synthetic products in a little more than two decades had a devastating effect on the environment and on limited resources.

[4] *Environment*, Vol. 13, No. 3, p. 4.

If the new technologies had not developed, Commoner maintains, the increase in population and affluence would have had a much smaller impact on men and the environment. "Social choices with regard to productive technology are inescapable in resolving the environmental crisis," he wrote.[5]

To halt the excess pollution and the inroads the new products are making on raw materials, Commoner and others argue that the nation must choose between nonpolluting and nondepleting natural products and the pollution-creating and costly—in their drain on resources—synthetics. Thus the choice might be posed between cotton and nylon; between the use of natural defenses in the fields and pesticides; between soap and detergents; between natural and synthetic rubber. Going further, the country might be compelled to shift its support from polluting automobiles to relatively clean mass transit, or from superhighways to railroads that disturb the countryside much less. Essentially, the choice would lie between new technologies and the older, less environmentally disturbing ones.

There is disagreement with the thesis that the new technologies rather than population are the "prime" cause of the deterioration of the environment. There is virtually no disagreement that they are an important cause of the destruction.

The first world systems study of the earth's predicament traced that predicament to the complex interaction and unlimited growth of population, industry, and pollution, and to the unlimited growth of demand for food and more resources. The resultant disruption and depletion were found to threaten the survival of the earth as a home for man, and a crash, meaning degradation, starvation, exhaustion and collapse, was predicted in not much more than a hundred years.

The studies were sponsored by the Club of Rome, a

5 Ibid., p. 19.

small, private, international group alarmed at developing conditions. The first of the studies was made at the Massachusetts Institute of Technology, under the direction of a team headed by Donella and Dennis Meadows. Using systems analysis, the computer, and the scientific method, the MIT team projected what might be expected if the present exponential, indefinite growth continues on this finite earth. The study was published as *The Limits to Growth*,[6] and the equations developed indicate that the earth's whole dire situation is the result of the interaction of five major factors—population, food production, industrialization, pollution, and consumption of nonrenewable natural resources—and not of any one, two, or three. The new technologies that Commoner singled out were appraised in *The Limits to Growth* as a cause but not alone the primary source of trouble. All five entered into the maelstrom, and one could not be separated from the others, the Meadowses maintained.

If any one of the five should continue at the present unrestricted rate of growth, it was shown that it could set off the predicted crash. A continuation of the present rate of world population increase would require a huge increase in food, industrial output, and services. Under this circumstance, the computers indicated, the crash would be precipitated by the depletion of resources. The scarcer resources became and the more depleted the mines, the higher prices would rise. More capital would be needed to mine or extract natural resources and less capital would be left for industrial growth. If the industrial base collapsed, it would take with it the service and agricultural systems dependent on industrial inputs. This would include fertilizers, pesticides, computers, and energy for mechanization. With population continuing to rise, the death rate would be driven upward by lack of food and health services.

[6] Donella H. Meadows et al., *The Limits to Growth: A Report for the Club of Rome's Project on the Predicament of Mankind* (New York: Universe Books, 1972).

Critics of *The Limits to Growth* dismiss this projection as unfounded. They charge that it does not take into account the development of new technologies and discoveries that would supply human needs if certain natural resources were practically exhausted. They point to the developments of the past—the electricity that replaced water and wind power, the automobile that took over for the horse, the synthetic rubber that replaced the natural, and many more. If the fossil fuels are no longer available for power, many are confident that they can be replaced with nuclear energy or solar power.

The Meadows team, in their awareness of this argument, examined what would happen if new technologies doubled the estimated reserves of resources. In such a case, they found, an increase in pollution would bring on collapse. The death rate would rise as pollution killed crops and the food supply dropped.

Again the critics object that the modern world has shown some ingenuity and some resolve in controlling pollution. The Meadowses therefore assumed that strict pollution controls would be instituted at whatever the cost. In such an event, the computers showed that the collapse would come from food shortages. The air and water might be good, but to feed the growing population more land would be brought into industrial and urban use, and the limits of arable land would be reached. Industrial expenditure would have to be diverted into food production. As this was done, industrial output would fall, and food production would also be affected. Life would sink to the subsistence level, and once again there would be collapse.

Other authorities argue that with full employment of technology even the earth's known resources could yield enough food "to chase hunger from the earth." The "green revolution" is cited as proof of what might be done. Though green revolution methods require up to 50 percent more fertilizer and pesticides, the Meadowses assumed that agricultural output per hectare might be doubled by such an effort. This, however, would require a quadrupling of industrial production—and a pollution crisis would stop growth, they found.

If increased food production were combined with the halting of population growth—populations held at about present levels—and pollution were controlled, the Meadows calculated that the earth would still be brought down by a triple crisis. Overuse of land would lead to a drop in food production, depletion of resources, and a final rise in pollution.

"The application of technological solutions alone prolongs the period of population and industrial growth, but it does not remove the ultimate limits to that growth," the Meadowses said.

Untrammeled economic growth has been the American and the world goal. Each year is expected to show a rise in output and profits, and the more than doubling of the gross national product in the postwar period testified to the achievement of that aim in the United States. Only recently in the face of the environment crisis has unlimited growth been called into question. John Kenneth Galbraith is one who has proposed that limits be set on the amount of production that the nation needs and that production be adjusted or cut back accordingly.

The Ehrlichs and others have also emphasized that a rising gross national product does not show the depletion of natural resources or the decline in the stability of the environmental systems on which life depends. They are advocating that quality of life be made a part of the measurement, and the goal.

The debate over how much growth can be tolerated has ceased being academic or theoretical. Cities and states that only a decade ago were struggling for new factories and more population are now sometimes rejecting both the goal and large commercial projects that would ruin the environment or pile on other public costs.

In Florida, for example, Governor Reubin Askew announced: "We are tightening up our restrictions for commercial development. We want to make sure that we are not simply attempting to grow for growth's sake."

Delaware refused a $360 million chemical complex.

The city of San Francisco turned down several proposed skyscrapers. They would have further blotted out the view of the bay—a view already partly destroyed by an elevated highway. Extension of the highway was stopped, and a proposal to prohibit all skyscrapers unless authorized by the voters received substantial support in a public election.

Roger Boas, a member of the San Francisco Board of Supervisors, told *Forbes* Magazine: "Developers have been too greedy in the past. . . . People across the state and across the nation feel they've got to do something to stop this promiscuous growth."

Even without further growth, our present numbers and technology are bearing down adversely on the physical and social well-being of millions. More than three-fourths of the people of the United States live in large urban and industrial areas where the air is no longer the original pure mixture of oxygen and nitrogen the lungs are adapted to breathe. It is loaded with soots and chemicals that are foreign to human tissues and that now seem to be highly injurious to the lungs and the whole physical makeup of humans.

At the extreme, air pollution can cause death. One December day in 1930 a thick, choking fog settled down in a heavily industrialized area of Belgium's Meuse valley. For three days it did not lift, and by the time it lightened hundreds had become ill and sixty were dead—far more than the normal number of deaths.

Another of the deadly, smoky fogs enclosed the small steel and chemical town of Donora, Pennsylvania, on October 26, 1948. Nearly half of the 12,300 residents became ill, and before the polluted air dissipated, 20 had died. When a shocked nation rushed in experts to determine what had happened, it was found that the haze was heavy with zinc sulfate, sulfur dioxide, and other pollutants. Ten years later a follow-up study disclosed that the Donora people who had been acutely ill during the fog had a higher rate of sickness and died earlier than the average for all the townspeople.

A few years later, in 1952, another great black fog settled down on London. In the first twenty-four hours people began to die in unusual numbers, and in the four days the fog gripped the city there were 4,000 more deaths than would have been expected under normal conditions.

Many of those who died in these disasters were suffering from heart or lung ailments, and many were old. They were, therefore, susceptible, but their deaths apparently were hastened because the chemical-saturated smogs that caused the disasters were held close to the ground by what is called a "temperature inversion," or a large layer of warm air above a layer of colder air. Ordinarily the pollutants rise and are dissipated over a wider area, but even when the winds are brisk, part of the 140 million tons of pollutants discharged from this nation's smokestacks and machines swirl through the first seven feet of the air, the air we ordinarily breathe.

Carbon monoxide in these emissions combines with hemoglobin in the red cells of the blood of all breathers. There it displaces some of the oxygen the hemoglobin usually transports to the cells. When the oxygen that enables the cells to function is reduced, the heart must pump harder to make up the deficiency. The extra effort may be too much for those with heart or lung disease.

Investigations have shown that when an individual lives for eight hours in an atmosphere containing 80 parts per million of carbon monoxide, the oxygen-carrying capacity of the circulatory system is reduced about 15 percent—which is just about the equivalent of the loss of a pint of blood. Occasionally, in a bad and prolonged traffic jam, the carbon monoxide content of the air may rise to 400 parts per million. In such concentrations symptoms of acute poisoning sometimes occur —headaches, loss of vision, nausea, pain, and convulsions. In a few extreme cases, unconsciousness and death have followed.

In Great Britain, where studies have been made of humans living in the polluted air of districts where there has been no cleanup, the incidence of bronchitis is high. Dr. Donald Reed of the London School of Hygiene found more respiratory disease in men who worked outdoors in dirty air districts than in those working outdoors in cleaner air districts. Men over fifty were particularly affected, and especially if they were smokers.

Physicians are not willing to say that bronchitis is caused by air pollution—it also occurred long before the air became bad. But statistics report more of it in the cities where the air is black. Contrariwise, bronchitis sufferers living in areas where Great Britain has reduced pollution said in response to an inquiry that they have fewer attacks. Each day they marked cards "better" or "worse," and while of course the judgments were subjective, hospital admission for disease associated with air pollution did decline in the cleaner districts.

Emphysema was once a relatively rare disease. In recent decades, however, it has become common, and is now the fastest-growing cause of death in the United States. In this disease, millions of the little air sacs of the lungs begin to fuse together, and the lungs become less capable of supplying oxygen. Smoking is believed to be one of the causes of emphysema, but air pollution, in medical opinion, aggravates it, if it is not indeed a partial cause.

The incidence of lung cancer has increased a hundred times in the last sixty years. Medical opinion now widely holds that a principal cause is smoking, but the studies indicate that the effects of smoking are intensified in the polluted air of cities.

Both new technology industries and old industries and

machines make extensive use of the heavy metals, nearly all of which are highly toxic to man and other living things. In the past little attention was given to the possibility that in one way or another some of the toxic metallic wastes might make their way into the food chains. Wastes were generally dumped into the waters or onto the ground, under the assumption that they would be dissipated before they could cause any harm. The case of mercury has lately jarred this complacency, and it is increasingly recognized that man's careless or economic actions may poison man.

In the 1950s fishermen and their families living around Minamata Bay in Japan became mysteriously ill. Their muscles and sight weakened, some lapsed into unconsciousness, and fifty-two died. Even more mysteriously, the cats and sea birds in the area seemed to suffer from the same illness. All, though, had one thing in common—men, cats, and sea birds lived on fish caught in the bay.

When an investigation was made it was found that all were suffering from mercury poisoning. Industries around the shore had been dumping mercury-containing wastes into the waters of the bay.

Other reports of mercury poisoning began to come in. Since 1960, 450 persons in several countries have become seriously ill, and some have died. In addition, three children in a New Mexico family sustained serious brain damage after eating pork from animals that had been fed on mercury-treated grain.

Mercury has been used in thermometers and in other ways for centuries. Since 1900, however, usage has increased steeply in paper and electrical industries and in fungicides. To meet the demand, production is now about 9,000 metric tons a year. It is thought that about one-third of this mercury is lost to the environment through smokestacks or by dumping on the land or in water.

In the air, in the form of vapor, mercury is highly toxic,

but the amounts involved are too small to be noticeable. In water, elemental mercury is not soluble, and it settles to the bottom. For a long time it was believed to be locked in mud as mercuric oxide. On the muddy bottoms of bays and lakes, however, bacteria sometimes convert the relatively insoluble mercury into soluble methyl mercury, a deadly poison. The methyl mercury is taken up by algae. The algae are eaten by little fish, and the little fish by larger fish, and at Minamata Bay the larger fish were consumed by the fishermen. Each consumer in turn concentrated more of the poison.

When the same methyl mercury was found in high concentrations in the fish of Lake Saint Clair along the United States–Canadian border, Canada acted decisively to stop the dumping of mercury in the area. The United States joined the effort, and later announced that mercury in the lake had been reduced by 86 percent. Policing, though, will be difficult, and mercury already in the water may continue to be methylated by fish and bacteria for fifty years or more.

Concentrations of mercury were also found in swordfish and other ocean fish. Its presence there seemed to be natural. Leonard J. Goldwater of Duke University, who made some of the mercury studies, concluded that the natural cycle of mercury through the atmosphere, biosphere, lithosphere, and hydrosphere disperses it so widely that it poses no hazard to life. The problem arises in man's alteration of the natural distribution.

"In the case of mercury," said Goldwater, "as in all other aspects of our environment, the wisest course is to try to understand and to maintain the balance of nature in which life on our planet has thrived."

Lead, another of the heavy metals, has caused poisoning throughout history. Heavy concentrations have been found in the bones of wealthy Romans; their wine vessels were lined with the metal. Since 1924, when lead was added to gasoline to reduce the "knock" in automobile engines, usage has increased from 1 million pounds a year to 700 million. Another 50 million pounds a year is added to aviation gasoline. And tests indicate that about 75 percent of this additive lead ends in the atmosphere.

Anyone walking the streets of a city or riding on a traffic-filled street, therefore, breathes in some of this lead along with other exhausts from all the running motors. By 1968 the average concentration of lead in the blood of certain city dwellers had reached 0.25 parts per million. There is a dispute over whether this level should be considered dangerous.

Many of the thousands of new chemicals are also dangerous or poisonous to man, as well as to other living things. DDT is one of the best known and most threatening of them to man, for man stands at the top of the DDT chain. We eat the fish that eat other fish that eat the plankton that absorbs DDT. And at each stage each consumer concentrates the pesticide.[7]

The milk of nursing mothers in the United States is now reported to contain from 0.1 to 0.3 parts per million of DDT, or two to six times the amount allowed in milk for commercial sale. Other reports indicate that in Sweden breastfed babies now get 70 percent more than the "acceptable" amount of DDT, and British and American babies, about ten times the permissible amount of dieldrin, a more toxic chemical than DDT. Some babies in western Australia are said to be exposed to as much as thirty times the "maximally acceptable" amount of dieldrin.

No one now knows what the effect will be of the poison the infant drinks with his mother's milk. Only time perhaps will tell. Some studies indicate that a maximum concentration has been reached by humans, probably of about 7 to 12 parts per million. Part of the difficulty in determining what the DDT group does to health is that human beings are now exposed to such a wide range of poisons that it is almost im-

[7] See Chapter 3, pp. 37-39; Chapter 4, pp. 63-67.

possible to separate out the effects chargeable to any one. The allegation has been made, however, that dieldrin may be involved in cirrhosis of the liver and benzene hexachloride in cancer of the liver.

Autopsies have disclosed a correlation between DDT levels in human fat and death. DDT and its breakdown products were higher in individuals who died of cerebral hemorrhage and various kinds of cancer than in patients who died of infectious diseases. Also, the histories of those in the study group showed higher DDT concentration among heavy users of pesticides.

Other studies, indicating that DDT is "safe" in human tissues, have been attacked as inadequate. Evidence is also developing that DDT may affect behavior. A seal herd off the California coast has for a number of years been exposed to an unusually heavy load of DDT from the extensive spraying of coastal vegetable fields. When the pups were born in 1970, some of the mothers behaved erratically. Instead of bringing pups back to the herd when they strayed away, these mothers nosed and pushed them in the wrong way until the pups died. The carcasses were counted on the beach.

The verdict on DDT in mammals and humans is not yet in, but the evidence is strongly persuasive. In 1972 the United States Environmental Protection Administration (EPA) ordered an almost complete ban on the use of DDT in this country. After December 31, 1972, all use of DDT was prohibited, except for public health purposes and for three minor crops—green peppers, onions, and sweet potatoes in storage—where effective alternatives were found to be unavailable.

William D. Ruckelshaus, administrator of the agency, said in a forty-page decision:

I am convinced by a preponderance of the evidence that, once used, DDT is an uncontrollable, durable chemical that persists in the aquatic and terrestrial environments.

The evidence of records showing storage in man and magnification in the food chain is a warning to the prudent that man may be exposing himself to a substance that may ultimately have a serious effect on his health.

The government order did not prohibit the export of DDT to other countries.

NOISE

Noise is another pollutant that may have to be taken into consideration in its effect on human health and well-being. In all cities, and particularly in the metropolis, noise is strident and constant. Pile drivers pound, sirens shriek, brakes screech, and planes roar overhead. There is a continuous background roar, punctuated by shrill outbursts of many kinds. If the threshold level of hearing is set at the number of 1, heavy automobile traffic or jet aircraft overhead have a noise level of 100.

The hearing of some individuals has been impaired by constant exposure to high noise levels, and even lower levels may interfere with sleep and produce fatigue. Whether the noise of cities is generally deafening is uncertain, but some evidence indicates that it may be. Many cities have enacted noise reduction ordinances in a effort to reduce noise pollution.

MAN'S VIEW OF HIMSELF AND NATURE

What men do about their world depends upon what they think about themselves in relation to it. In the end that is the determinant. These basic concepts tend to be ancient ones, and most frequently religious ones. Though this is the twentieth century and the age of science, many of the ancient attitudes toward nature are almost unchanged. Many people believe implicitly that dominion over nature is theirs. We split the atom, but we behave as though *The Origin of Species* had never been written. We are supplied with the figures on world population, but we act as though we must limitlessly obey the biblical stricture to "be fruitful and multiply."

Man's effect on the environment at first was slight. His numbers were few, and the waters, forests, and plains seemed endless. Survival, not conservation, was the problem. Whether any one man could find enough to eat and surmount the dangers and hardships of nature was questionable. Early men feared mysterious nature and its outbreaks. In those days nature had to be appeased; it was to be revered. It was holy.

It is believed that primitive men surviving today display many of the feelings and attitudes of early men. A group of Bushmen hunters pursued a giraffe for more than three days before they could bring it down with their short bows and arrows. They were hungry and weary. But before any meat was eaten they held a religious ceremony. A great light had left the world, and this had to be acknowledged. They killed, for man had to eat, but the killing was done with ceremony and reverence.

Throughout antiquity every tree, every stream, and every mountain had its own guardian spirit. Before a tree was cut or a stone was removed from a mountain the spirit had to be placated. Nature was still both feared and revered, and any changes were made carefully. This spirit, this attitude toward nature carried over into many later societies. The American Indian, for example, regarded the land as a trust, never to be misused or molested. Sometimes, if a few rocks were gathered to build a campsite, they were restored to their rightful places when the camp was abandoned.

The first agriculture, when men first began to exert more than a passing impact on nature, was also regarded as a rite. In the beginning the planting was done by women, and at a time believed to be acceptable to the gods. "The bounteous earth bore first for them of her own will, in plenty and without stint," sang Hesiod. Men continued to stand in awe of the bounty that was somehow given to them, and much of this feeling was never to die. Prayers of supplication, for nature's yield and for safety, and prayers of thanksgiving have remained a serious, essential part of man's life.

An invention then began to change some of the old feeling about the earth. It was the knife plow, or plowshare. The iron plow that had come into use did little more than scratch the surface of the land, while the new plow sliced the sod and turned it under. Nevertheless, the sod was thick in some of the lands to which the farmers spread, and they needed oxen to pull the plow. This forced them to pool their animals and their holdings. For the first time the distribution of land was based, not on the needs of one family, but on the capacity of a power machine to till and command the earth. As Lynn White, Jr., pointed out in his lecture "The Historical Roots of Our Ecologic Crises," man's relation to the soil was profoundly changed.

"Formerly man had been part of nature," he said. "Now he was the exploiter of nature." [8]

A feeling of power over nature began to develop, strengthening as the Romans conquered much of the known world. It seemed to the conquerors that their progress came through the exploitation of this vast territory. Cicero in *De Officis* clearly defined the conviction:

Those very things which we have called inanimate are for the most part themselves produced by man's labours; we should not have them without the application of manual labour and skill nor could we enjoy them without the intervention of man. And so with many other things: for without man's industry there could have been no provisions for health, no navigation, no agriculture, no in-gathering or storing of the fruits of the field or

8 *Science*, Vol. 155, p. 1205.

other kinds of produce. Then too there would surely be no exportation of our superfluous commodities or importation of those we lack, did not men perform these services. By the same process of reasoning, without the labour of man's hands, the stone needful for our use would not be quarried from the earth nor would "iron, copper, gold, and silver hidden far within" be mined.

Cicero continued his list. Without the labor of man and the use of the earth there would be no shelter, no arts, no medicine, no cities, no social organization, no law.

"Upon these institutions," Cicero added, "followed a more humane spirit and consideration for others, with the result that life was better supplied with all it requires, and by giving and receiving, by mutual exchange of commodities and conveniences we succeeded in meeting all our wants."

Seneca expressed much of the same feeling, although with grave doubt of some of the excesses of exploitation. He wrote:

Nature was not so hostile to man that, when she gave all the other animals an easy role in life, she made it impossible for him alone to live without all these artifices. None of these was imposed upon us by her. . . . Houses, shelter, creature comforts, food, and all that has now become the source of vast trouble, were ready at hand, free to all, and obtainable for trifling pains. For the limit everywhere corresponded to the need; it is we that have made all other things valuable, we that have caused them to be sought by extensive and manifold devices. Nature suffices for what she demands. Luxury has turned her back upon nature; each day she expands herself, in all the ages she has been gathering strength. . . .

Christianity was also triumphing over paganism, and man's view of the world was being further changed. Where the ancient religions had espoused a cyclical concept of time and life and gave only a mythical account of the earth's beginnings, Christianity proposed a grand creation by a God who brought earth, man, plants, and all living things into sudden being, or so the biblical words were generally interpreted.

Man's relations to this creation were clearly set forth. In Genesis it was declared that men have "dominion over the fish of the sea and over the birds of the air and over every living thing that moves upon the earth." Man was made of clay, but the clay was shaped in God's image. Man, then, was not merely another part of nature.

As early as the second century Church authorities were insisting, further, that it was God's will that man should exploit nature for his proper ends. The sacred trees and groves of antiquity were deliberately cut to banish "idolatrous" paganism. Many men concluded that they could be indifferent to the feelings of natural objects, and in time even a little girl dragging a sparrow by a string replied in the following way to protests that she was being cruel to the bird: *"Non est Christiano."*

The dominion concept was reinforced during the Middle Ages, as all eyes turned to God and life on earth was regarded as a mere passage to the hereafter. The spirit of the times and the attitude toward nature emerged in the painting of the period. Nearly all of the subjects were religious, and if a natural object such as a dove or a pelican or a lily was used, it was only as a symbol of some religious concept or figure.

The literal biblical words affirming man's dominion over nature sank deep into the consciousness of men. They became, as Alfred Whitehead said, the form of thought, and "like the air we breathe, such a form is so translucent and so pervading and so seemingly necessary that only by extreme effort can we become aware of it." As embedded as the concept was, it was further supported, on rational and philosophical grounds, by such sixteenth- and seventeenth-century leaders of thought as Bacon, Descartes, and Leibniz.

In his *Novum Organum* Francis Bacon upheld the idea of dominion, though with some caution:

If a man endeavour to establish and extend the power and dominion of the human race itself over the universe, his ambition (if ambition it can be called) is without doubt both a more wholesome thing and a more noble than the other two. Now the empire of man over things depends wholly on the arts and sciences. For we cannot command nature except by obeying her.

Leibniz, in an exalted search for the good, proposed to attain it and make the earth more perfect by bringing it further under the guiding hand of man. The cultivation of the earth was equated with perfecting it. He wrote:

> To realize in its completeness the universal beauty and perfection of the works of God, we must recognize a certain perpetual and very free progress of the whole universe, such that it is always going forward to greater improvement. So even now a great part of our earth has received cultivation [culture] and will receive it more and more. . . .[9]

Though perfection was unobtainable, the progress made was impressive to the Western man of the eighteenth and nineteenth centuries. He had spread all around the earth, and he was in control. There seemed to be few limits to what he could expect to accomplish.

"The struggle with and the control over nature were ways of depicting the progress of civilization," said Clarence J. Glacken of the University of California, Berkeley, in "Man Against Nature: An Outmoded Concept."

In its material aspects civilization meant this—the purposive changes in nature, the overcoming of natural obstacles by bridges, drainage, roads, and later by railroads, air and sea routes.

There was little awareness or study of the age-old cumulative but unnoticed effects of man's activities in changing the environment. There is ample evidence of local awareness, but such

9 Gottfried Leibniz, *The Monadology and Other Philosophical Writings,* translated with an introduction and notes by Robert Latta (Oxford: The Clarendon Press, 1896), pp. 350–51.

knowledge was not widely diffused.[10]

Many voices proclaimed this increasing dominance over nature, and hence progress. Walt Whitman in his "Song of the Redwood Tree" did not mourn the loss of the forest giants to the axes of man, for he found them yielding to a "superior race," and willingly giving way to civilization.

An anonymous poet of some years earlier—1692—had celebrated, with no doubts, the felling of the forests:

> In such a Wilderness . . .
> When we began to clear the Land . . .
> Then with the Axe, with Might and Strength
> The trees so thick and strong . . .
> We laid them all along . . .
> Which we with Fire, most furiously
> To ashes did confound.

The cities with their monuments were further testimony to what man had done. Samuel Woodworth, in 1827, was speaking of the Grand Canal, but the implications were much wider:

> For ours is a boast unexampled in story,
> Unequalled in splendor, unrivalled in grace,
> A conquest that gains us a permanent glory,
> The triumph of science o'er matter and space!

The same sense of triumph was expreseed by Jacob Bigelow, in an Inaugural Address delivered in 1816: "The ingenuity of mankind . . . has never forgotten how to subjugate the elements to its will, and to reduce all natural agents into ministers of its pleasure and power."

As machinery and the industrial age increased the power of man over nature and some of the drudgery of scraping an

---

[10] In *The Environmental Crisis: Man's Struggle to Live with Himself,* edited by Harold W. Helfrich, Jr. (New Haven and New London: Yale University Press, 1970), p. 131.

unaided living from the soil was relieved, the belief that man was the master of nature grew all the stronger.

"The head saves the hands," Theodore Parker wrote in 1841. "It invents machines which doing the work of many hands, set free a large portion of leisure time from slavery to the elements. The brute forces of nature lie waiting man's command, and ready to serve him."

Though man was successfully subduing nature, nature in another way was not unappreciated. Great gardens were laid out in France, England, and other countries. The French gardens, of which Versailles was the archetype, were ordered, symmetrical, patterned. The English used nature somewhat more naturally, but even the big trees, lawns, and bright flower beds were arranged in chosen places and vistas. Here, too, nature was manipulated.

The development of science in its turn increased man's assurance and his feeling of control over nature. As men began to understand some of the laws that underlay the workings of nature, some of the ancient mysteries, and the awe of them, began to vanish.

The ingrained and vaunted belief that man could control nature—and should—was challenged from time to time. Within the Roman Catholic Church, Saint Francis of Assisi preached a different doctrine, one based on the humility of man as an individual and as a species. All men and all living things, he taught, are brothers in God's creation. Man, then, was not the master. To Saint Francis the ant was created not to warn man against laziness, as the natural philosophers of the Church insisted, but as a living thing, glorifying the Creator in its own way, much as man did in his. The concept could have changed the religious belief in domination, but it failed to do so. It did not prevail.

Charles Darwin brought in another revolutionary concept. In his work he pictured all living things as part of nature. All were related, and all were interdependent in the great webs that extended into known and unknown parts of nature. And balance was achieved through these relationships. Men and nature were one.

George Perkins Marsh, a naturalist as well as the first American ambassador to the kingdom of Italy, in 1861, was one of the first to warn against the consequences of disrupting nature. In his book *Man and Nature* he traced the cumulative changes man had made even then—deforestation, reclamation, drainage, and "the dangers of imprudence and the necessity for caution in all operations which on a large scale interfere with the spontaneous arrangements of the organic and inorganic world." Marsh feared that the damage might be irreversible and that it would end with the planet's becoming uninhabitable.

Some painters and artists also broke away from the man-the-master concept. In the eighteenth and nineteenth centuries Turner and Constable were two who celebrated nature. Any men, women, or children in their paintings fitted gently into the expanse of trees and clouds.

The East, too, took a different approach. From the beginning the men of the East, like other early men, saw deities in every form of nature. The gods assumed the form of birds, monkeys, elephants, and other animals in order to manifest themselves to man. There was also a deep sense of continuum from man to nature. In the endless cycle of life man might be born again as an animal. Some sects turned to vegetarianism out of their reverence for nature, and certain monks would go to any length to avoid stepping on an ant; all life was sacred. All sentient beings were a part of nature and held in respect because of that oneness.

The Chinese and Japanese, too, looked upon themselves as a part of nature and not as above it. Nature was reverenced and treasured, as Oriental painting has testified through many centuries. In the twelfth century the critic Han Cho wrote: "Painting . . . stands in subtle accord with the creative process of nature." And painting was to convey the forces of nature to man—"the haze, mist and haunting spirits of the countryside that are what human nature seeks and rarely finds." It was never simply man against nature.

The West, despite the dissenters and the East, continued generally to believe in the biblical dominion of man over nature and to equate mastery with progress. Even those who did not consider their attitude as Christian, or as an outgrowth of ancient Christian theology, unconsciously accepted it. As Lynn White has said:

> The fact that most people do not think of these attitudes as Christian is irrelevant. No set of basic values has been accepted in our society to replace that of Christianity [on nature]. We must re-think and re-feel our nature and destiny. . . . Since the roots of our trouble are so largely religious, the remedy must be essentially religious.[11]

Some rethinking and refeeling has come in the modern scholarly reexamination of biblical texts and antecedents. The new studies of the Bible and its sources have shown that Jesus said very little about nature. Nature was brought in principally by implication or as a fact or judgment of God's goodness, in such references as "the rain falls alike on the just and unjust," or "the lilies of the fields." Nature was also used as a contrast to man's pride, with the natural order as an analogue to the human order.

Scholars have traced most of the prevalent Christian doctrine of nature to Paul. Paul, with great distrust of the world, preached that man must "transcend" nature. The nature of which he spoke, however, was primarily the nature of man; it was the sins and the natural functions of man that Paul sought to overcome or repress. Nevertheless, the scholars find, the Pauline position was extended into the theological justification for the concept of man's superiority over nature.

11 *Science*, Vol. 155, p. 1207.

New translations of the word "dominion" used in Genesis are also giving it more the cast of stewardship than of superiority.

Nevertheless, the original translation and a farfetched extension of the doctrine of one prophet entered influentially into the concept of nature that was to influence and direct the actions of much of the Western world for more than a thousand years—and that made Western man dangerously oblivious to the reality that he is a part of nature.

Though Western man, with little compunction, turned nature to his own uses, the voices of warning distinguishing between use and abuse were faint until the middle of the twentieth century. To the individual the world seemed large and inexhaustible, and any one act of pollution apparently mattered little, so the warnings that unlimited exploitation might not be possible came as a shock and surprise. Even the warnings are as yet not backed up with many specifics of where the breaking points and crises might come.

As Dr. Cecelia Tichi of Boston University has written:

Over the last millennium we have shifted from a fear of the threatening vagaries of nature to the fear of unforeseen results of our own wretched planning. The fertility rites acknowledging gods of fertility, growth and harvest were carried on as much from fear of scant harvest and potential starvation as from affection for those gods. And the demons and dragons of the Bible, coupled with the Anglo-Saxon folklore about a menagerie of fearsome beasts, strongly suggest deep human fears about the natural world. Now, of course, it is human technology that has replaced the bestiary as a source of fear, simultaneously calling into question our notions about human progress.

The pristine world of many dreams will never again exist, but outside of his dreams man might not want it. It was a hard and fearsome world, to which we would not willingly return. But we might desire a renewable world, where overcrowding and abuse do not destroy the systems on which all of life depends. If man and nature are one, and man sees himself as the steward, not the master, the adaptability of the human race indicates that man and nature may survive together.

Chapter 6

IMPERATIVES AND ALTERNATIVES

At a moment of crisis, a sage once assured his troubled country: "That passed, this also may."

This, however, will not. The stark truth that man must at last face is that the problems of population, pollution, and scarcity will not pass away. They cannot.

The earth cannot be enlarged. This is a finite, limited sphere on which we live.

Population will increase "to any number," unless it is checked, voluntarily or by catastrophe.

Man is a part of nature. If he destroys his environment to accommodate his numbers and wants, he destroys himself. The two are inextricably linked.

These are the imperatives. They always have been, it is true, but until recent times men were not pressing against the limits of the globe. Now they are. In the same way, it is inevitable that a glass will overflow if enough drops of water are dripped into it. As long as it is only half full there is no problem. But if it is nearly full and the dripping continues, over-flow is inevitable. The problem will not pass. It cannot.

POPULATION CONTROL

Even if we learn to control our exploding population today we will be left forever with a population that will have to be carefully watched and managed.

Even if we prevent or reverse pollution of the environment, we will not only retain *forever* the ability to pollute air, land, and water, but probably will develop new means of pollution which cannot be imagined today.

Even if we learn to conserve those natural resources which are renewable, potential breakdown in conservation policies will *forever* threaten the exhaustion of the very substance of this planet.

We have entered a new world in which today's major problems are literally with us *forever*. We will have to change and adjust to these circumstances . . . to "seek nothing less than a basic reform of the way our society looks at problems and makes decisions" or suffer the consequences *forever*.[1]

That the earth cannot be enlarged is self-evident. The exploration of the moon and the probes of Venus have also established that the nearest planetary bodies are unsuitable for man. To support colonies on either or on space stations would preempt so much of the earth's resources that the earth would lose far more than it could gain. So shifting excess people to another planet is not a feasible alternative.

If we must live on this planet, and if this is a closed system, as it is except for the input of solar radiation, it follows that unlimited growth will have to be checked. Otherwise we are, as The Institute of Ecology warned, "on a collision course with nature."

To restrict runaway growth will clearly require:

[1] From a statement issued by the Field Museum.

A SLOWING DOWN or halt in the present rampant rate of population increase. We cannot continue adding 500 million in a decade and doubling every thirty-five years. If we continue at even these current rates of increase, the often uttered prophecies of doom must come true. Either man himself or some ultimate disaster will reduce populations.

THE USE AND REUSE or recycling of the earth's resources. No earth supply is inexhaustible, except for the sun's rays, and even the sunshine can be adversely affected by pollution. From the beginning, all the resources used by plants and animals were put back and thus used and reused through all the 3.5 billion years of life. Only with modern agriculture and production was the age-old system of use and return disrupted in part. Unless it is substantially resumed, man sooner or later will run out of food and other supplies necessary for the continuity of life.

A DRASTIC REDUCTION in the nondegradable, manmade chemicals and the poisonous heavy metals now being introduced into the atmosphere, the waters, the soil, and the biosphere—the realm of living things. Few bacteria and fungi have evolved to break down these manmade chemicals to their original and reusable elements; nor can these pollutants be easily removed. They do not stay where they are introduced, but spread to the ends of the earth and to every zone. DDT has been found in the penguins of the Antarctic.

THE PRESERVATION AND UP-BUILDING of diversity. When only one or a few species are present, nature is subject to drastic fluctuations. A single species or a single crop is susceptible to attack by pests, disease, or pollution and can easily succumb. When there are diverse species all are not vulnerable; some are certain to survive. If there is to be stability and continuity, and any hope of producing enough food for even the present populations, diversity must be preserved. It is an ecological imperative.

Can we meet these imperatives and alternatives? The

possibilities are increasingly being studied by science. The final answer is not known, but some procedures and policies are being developed to cope with this matter of life or death for a planet and the unique life that occupies it.

In the first place, can population be checked before people overwhelm the earth? Famine and disease have tended to keep numbers down through most of man's history. When this decimation was reduced, principally in the last century, population skyrocketed to the present 3.5 billion, and unless restraints are imposed, the earth's population will go on "to any number."

Will man impose such restraints? Can he? Through a variety of taboos some primitive societies have kept their numbers stable. The most advanced industrial societies have also multiplied at a somewhat slower rate than nonindustrial societies. But the effect of this slowing among the primitives and the industrialized city dwellers has been negligible. The world increase has still been formidable.

Through the middle part of this century, groups seeking to promote family welfare and to protect women from excessive childbearing developed modern methods of birth control, and toward the end of the 1960s the United States also began to lift some of its former restrictions on abortion. The birthrate in this country fell from 25.3 to 17.8, or about one-fourth, in the decade from 1959 to 1968. How much of the decline was attributable to birth control and abortion was not certain; it might also have been affected by the smaller number of women in the fifteen- to forty-four-year-old childbearing age group. Demographers hesitated to predict any lasting slowdown in the population growth rate of the United States, though the rate continued to drop in the early 1970s. However, to stabilize the population of this country at the present level of 200 million, the birthrate would have to drop to 13 per thousand.

Organizations are forming to achieve a stable population.

One group, called ZPG, for "zero population growth," advocates only two children per family, a rate that in time would hold the population to about its present size. Legislative measures have also been proposed to persuade the people of the United States to accept a population limitation. Prominent among them are tax changes to favor the small rather than the large family.

Japan was one of the first nations to embark on population control. Abortion was legalized, and the birthrate subsequently dropped from 28.3 to 17.2. Although Japan halted an increase that was threatening to swamp the small group of islands, the Japanese population has still not been stabilized. It is increasing at about the same rate as that of the United States.

Burgeoning populations have created equally acute problems in other parts of the world. In the undeveloped countries, where mosquito and disease control dramatically reduced the deaths of infants and children, between 40 and 45 percent of the population is under fifteen years of age. Populations will explode again when the many young who have survived become parents.

Experience has shown that some means for controlling the runaway human population of the world are at hand, but also that all of the means used so far are not perfected. Nevertheless, birth control has been used on a wide enough scale, particularly in the United States, Europe, and Japan, to prove that it can reduce the birthrate. Even now, man could if he chose prevent his numbers from increasing to more than the earth can support. The question is, will he?

If the population of the world should double in the next three decades, The Institute of Ecology has estimated that to double the amount of food and other support for the 7 billion who would be alive in the early years of the next century would take 6.5 times as much fertilizer, 6.0 times as much pesticide, and 2.8 times as much power as we now have. Because of the greater difficulty of producing more, a doubling of output to match the doubling of the population would not be sufficient.

Said the Institute:

We recommend that every effort be devoted to ensuring that the world population stop growing at the earliest possible date. We recommend that an international agency be established. . . . In the absence of a world population policy it is impossible to formulate rational policies for the use of the world's resources.[2]

RECYCLING

Even with present levels of population, resources are being strained and drained, and in time the strains and drains are bound to increase. If the earth is unable to supply enough food and shelter, the first effect on people will be a lowering of the quality of living. Degradation would be certain to continue, and if relief was not found, extinction would ultimately result.

The strain and drain are relatively new. Until recent times most of what was taken out of the soil was returned. Natural plant communities took the substances essential for life from the soil and air and used them, with the energy of the sun, to manufacture organic materials, which in turn provided food for a long chain of herbivores and carnivores. Little was lost, and at various stages materials were returned to the soil or air or water. As plants and animals died, decay liberated the nutrients for plant reuse. It was and is a balanced cycle, one that could keep going indefinitely on an earth where the resources are limited.

Now that man is practicing industrial agriculture, the crops—the single species crops—take a tremendous weight of

[2] The Institute of Ecology Report of the Workshop on Global Ecological Problems, titled "Man in the Living Environment," 1971. Subsequent Institute quotations are from the same report.

nutrients from the soil. And these nutrients are not returned, as they are when wild plants die; unless they are replaced by man, future crops will not grow. The replacement comes in the form of fertilizers produced by our industrial system. Factory-made pesticides are also used to eliminate insect pests, and industrially produced fuel must be drawn upon to operate the industrially produced machinery that enables us to cultivate large fields.

When the output of these fields is shipped off to the cities to feed the millions there, neither do the nutrients go back to the soil. Vast quantities of them are carried off in sewage, and generally the sewage is dumped in the nearest river or lake or estuary. In the waters and the seas that the sewage eventually reaches, the nutrients are not only lost to the land cycles, they cause the water's aesthetic deterioration and add to the cost of purification.

Said The Institute of Ecology:

It is a gigantic one-way flow of elements from the earth and the air into the sea. The scale of the operation is far greater than anything previously known on the face of the earth. And this human phenomenon is in stark contrast with the natural communities of plants and animals which have been living in balance with their surroundings for thousands of years.

Since we cannot draw indefinitely from a limited supply without also returning, the Institute has called for international action to devise methods for returning the nutrients to the terrestrial cycle of biological production.

High on the list is the management of the land, for many of the decay processes that lead to recycling occur there.

In the grasslands of the world this would require the prevention of overgrazing and excessive dry land farming, as both destroy the plants that keep the soil stable with their extensive root systems and their reserves of food and water. Nomadic and rotational grazing are recommended. Such a system would supply man's needs, but if not overdone would maintain the grasslands for continuing use. The alternative—continuing misuse—would turn many of the remaining grasslands into wasteland dust bowls.

In the tropical lowlands, management would entail a slowing or halt in the destruction of the forest, the most productive of all ecosystems. The cutting of the forest quickly destroys the fertility of the soil. In the high temperatures and heavy rainfall of the tropics, the nutrients are almost immediately washed out of the soil and lost. Instead of unlimited settlement and intensive modern farming in the tropics, a return could be made to the shifting agriculture of the past, a slash-burn system that burns off the land and permits the growing of crops as long as the nutrients last, but then abandons the land and permits the regeneration of the forest.

In arid and semiarid regions, good management might make possible the use of more of the dry lands for grazing. However, overgrazing quickly breaks the delicate surface crust, known as "desert pavement," and lets the soil blow away. Oscillations in the scant rainfall produce large oscillations in the density of the plants, and this in itself tends to produce simplification and instability.

Management will also be necessary to maintain the diversity and stability of other areas of the earth.

The one-way outflow of nutrients from the land to the city to the sea can be changed by recapturing the nutrients in the sewage and returning them to the land in various ways. The sewage pouring into the waters is rich in organic matter and plant nutrients. Both have always been part of the natural cycles, and nature is adapted to making use of them. The ecologists urge the use of these nutrients on the land and in the growing of fish. Sewage effluent is already widely used in this way in many parts of the world.

In one experiment, algae were grown on sewage and the algae fed to oysters. The oysters filtered the water, and their droppings fed worms on the bottom. Fish ate the worms and seaweed used the nitrogen compounds excreted by the animals. Such a system reuses all the components of sewage effluent.

Another system used carbon dioxide from a power station chimney to stimulate the growth of algae. These algae were fed to clams growing in the heated waters emitted by the power station.

Strong international management would be expected to use many means to feed back the nutrients man now diverts from the natural system, and thus to make it possible for man to continue to live on the limited resources of this earth.

A clear distinction must be made between the organic and inorganic nutrients on which all life depends, and such manmade chemicals as the pesticides and the natural but poisonous heavy metals.

The former, in sewage or any other form, promote plant growth, though under the present system they may promote so much of it that the waters are filled with low-grade and rotting plant materials.

The manmade chemicals—the pesticides and industrial products—are either unusable by life or are toxic. The unusable ones are stored, usually in the fat or bones, and passed along to all animals that eat the original consumers. In sublethal amounts the chemicals may produce abnormal behavior, as in the DDT-saturated mother seals that were unable to round up stray pups in the traditional manner, causing the deaths of the pups. In larger amounts such chemicals can be lethal.

Man has no immunity to the new chemicals; he is no exception. Many of the chemicals now in the food chains—the meat and fish man eats and the milk he drinks—have been proved to be tumor inducing. Others are now known to produce genetic defects.

"It is essential that mankind make every possible effort

to keep these substances out of the living environment," the ecologists said.

The SCEP report proposed strict measures to reduce the volume of such compounds and either to neutralize them or shield men, plants, and animals from their effects. The report recommended concentrating remedial measures at the sources of pollution, or at other points of concentration.

The costs would be high, but the SCEP held that the job could be done. One way would be to change the laws to force industries discharging or making such chemicals to "internalize" the costs, or to pay for the damages as they now pay for any material used in their manufacture. If this should cause an industry to fail, and if its produce were important to the public good, public subsidy was suggested to meet the costs.

The Institute of Ecology concurred in the SCEP recommendations, and proposed in addition an international control commission. Such an international body would set safe standards for the manmade toxic materials and for others that are dangerous. It would also maintain a program of emission inspection for all global pollutants.

"We recommend that all nations adopt comprehensive regulations to control production and release of toxic materials within their borders," said the ecology report.

The restriction of pesticides would not mean a surrender to insect pests or weeds. On the contrary, it would be planned to save the world from new pests and perhaps ruin by pests no longer susceptible to the chemicals. Since the spraying of cotton with DDT started in 1943, almost every major cotton pest species has developed resistance, and other insects, relieved of their predators, have flared up until they have become pests.

Integrated control is recommended as an alternative to deluging everything with chemical poisons. A major element in it would consist of biological controls, or the use of natural predators of the pests. The case of the Cañete valley was previously cited. There the new pests that surged up after the spraying reverted to their former innocuousness when the spraying was halted; their predators were no longer being killed. The cotton production in the valley not only returned to 526 kilograms per hectare—it had been down to 300 when DDT was used—but costs of control were reduced. Cotton yield has since varied from 724 to 1,036 kilograms, the highest yields in the history of the valley.

Integrated control would also include habitat management and the spacing or timing of crops to break or interfere with the life cycles of the pests attacking the fields. In one area, cutting of alternate strips of alfalfa controlled the lygus bug pest by killing its nymphs and forcing the adults to migrate at a time when most of them would be killed.

The development of resistant strains, or plants with their own genetic resistance to insects, would form another part of integrated control. Experimental work toward this end is going forward in a number of universities.

The Institute of Ecology proposed a United Nations program to further training in the planning and management of environmental systems:

> Dependence on chemical pest control endangers the long-term dependability of our agricultural ecosystems, causes serious environmental degradation, and sometimes poses a health hazard. Every effort should be made to replace chemical pest control with other methods, and particularly with an integrated approach that will develop and use ecological diversity.
>
> We recommend that all pesticides suspected of causing environmental damage or hazards to human health be subject to international restrictions and that the use of such pesticides be banned by international agreement when alternatives are available.

If critical damage to life is to be averted, early and stern measures will also be required to control the dissemination of the heavy metals.

Both the SCEP and The Institute of Ecology call for the elimination of tetraethyl lead from gasoline; they emphasize as well that lead is a dwindling resource and that all possible measures should be taken to conserve it. The two groups also recommend the barring of all mercury-containing pesticides and the recouping of mercury before it is emitted by industrial smokestacks.

Cadmium, a by-product of zinc production that is extensively used in superphosphate fertilizers, should be monitored and carefully studied, the ecologists said. It makes its way through fish and meat, and also through the air, into human tissues. A recent study of human death has linked hypertension with concentration of cadmium in the kidneys, and a similar cause of death has been found in other animals.

Great caution is recommended in the use of the polychlorinated biphenyls. Like DDT they accumulate in the fat and are extremely persistent. The Institute of Ecology report recommended only the most necessary use and careful control of stacks to prevent emissions.

Warnings were also extended about arsenic, beryllium, nickel, and ozone. All of the dangerous heavy metals, as well as the manmade chemicals, should be brought under the control of an international pollution control commission, the Institute said. All nations were also urged to adopt their own regulations.

If men are to have food and a livable environment, action must be taken on another of the changes man has blindly made in nature. From the time the first grains and animals were domesticated, the many species of the wild were replaced with a few species deemed necessary or important to man. Almost none realized the danger that lay in the choice, and eventually millions of acres were planted with single crops. As modern agriculture came in, species diversity was further reduced in the interest of efficiency. "Weed" plants were killed with herbicides and "pest" species with insecticides and fungicides. Where there had been many, there now were fewer. And many other species were lost by the destruction of their habitats.

When there were many, species were held in bounds by their natural enemies. All were likely to be present in the diversified environment. Any runaway was soon halted, and there was stability. Forests, fields, and waters thus continued substantially as they were for century after century.

With only one crop and a few species of animals, the few may multiply to enormous numbers. Very little stands in their way. But if disease or a pest or drought strikes, the numbers are vulnerable. The whole can be lost, instead of only the few among the many. It tends to be an all-or-none gamble, with a built-in likelihood of loss.

In a world where the losses are becoming serious, the problem is how to maintain the diversity that remains and to restore it where it has been reduced. If a species is totally gone, as is the dodo, no restoration is possible.

To maintain diversity, a diversity of landscape must be kept. And this includes grasslands, lakes, forests, estuaries, rain forests, tundras, and all the varied zones of the world. Many species can live only in their own environment.

If settlements and expanding agriculture continue to encroach on the forests and arid lands, if cities sprawl further over the productive countryside, if more estuaries are filled in, and the dumping of more wastes increasingly pollutes the oceans, diversity will be the first loss. The second will be the stability of the affected environment. And the third will be resources for the future.

When man turns to biological controls to save agriculture and must seek out predators and parasites to control pest species, the needed insects may not be in existence if more of their habitat is destroyed. Without the great richness of species that originally existed, biological control will be severely handicapped or perhaps impossible.

Genetic variations in wild populations also serve as a potential resource for improving crops and animals. A wild species of plant better able to resist the drought may provide exactly the genes needed to produce a hardier and better domestic plant. Without such variations to draw upon, little could be done in breeding.

Some wild species may in the future be changed into a crop species or used for some other purpose of man. The domestication of the eland, for example, has been proposed; its development might make possible the opening of areas where cattle cannot live. And other species may be needed for medical requirements not now foreseeable. Thus diversity may be one of man's most valuable resources.

Certainly, preservation of a few chosen species will not be enough. However important for present or future, the whole diverse ecosystem may be even more essential for man. The whole, little-known framework of exchanges, interactions, and organization may be vital if there is to be enough stability for human life and well-being. An upset at one remote point could upset the whole or indispensable parts of the whole.

We also rely absolutely on complex, diverse natural systems to dispose of our wastes. Without bacteria, fungi, and other agents of decomposition, man would long since have starved or have been lost in his wastes; the recycling that returns the elements for reuse would have ended. Most of the decomposition occurs in the soil or on lake or sea bottom. If these systems failed to work, man would be almost totally unable to replace them, and there would be little hope for the human species. Thus mud—plain, lowly mud and the undisturbed places in which it accumulates—is seen again as one of the true indispensables.

But can this all-important diversity be maintained? Is diversity possible in a world where 41 percent of the land is either moderately or intensively managed by man, and where some of the original diversity has already disappeared? The managed land includes the millions of acres in argricultural use.

Ways of coping with the problem of diversity in the food-raising and managed areas have been devised. One is called "spacial heterogeneity," or "mosaic planting." Instead of planting endless fields of grain, it may be necessary to intersperse the grains with other plantings. The lesson of the balsam forest holds promise. When balsams are mixed into hardwoods, infestations of the spruce budworm are low, since insects that prey on the pest worm find refuge in the hardwoods. Thus stands of the conifers in hardwood forests have lower infestation rates than continuous balsam forests.

In a heterogeneous environment, different events occur at different times, often preventing drastic fluctuations of animal and plant populations over large areas. This mechanism was demonstrated in a population of turban shells under attack by starfish. Though the snails were wiped out at places, at other places along the shore, where events occurred at different times, they survived.

Studies of the diversity of the insect species of hedgerows, adjacent pastures, and bean fields showed that species diversity on vegetation and in the air varies in a consistent way over spatially heterogeneous vegetation. Diversity was greatest in and over the hedge, lowest in the pasture, and intermediate in the bean field. The hedgerow apparently enriched the insect species diversity above and well outward from both sides of the hedge.

"Spatial heterogeneity is probably the most fundamental and important aspect of the relation between diversity and stability," said The Institute of Ecology report.

Variety within a population also aids in the maintenance of general diversity. If there are individuals of different ages and different genetic endowment, a population is more likely to survive attack by disease or predators—a fact also demonstrated in the balsam forest. Mature trees were more susceptible to infestation by the budworm than were young trees. Where young and old trees were mixed, there was less infestation, and the stands were more stable than in places where all the trees were of one age.

If agriculture is to provide food for the future and if the earth is to continue to support human life, all of the ecological imperatives must be taken into account:

Agriculture and other systems must be based on recycling most of the nutrients and using solar energy.

Extreme caution must guide the use of toxic, long-lasting pesticides and herbicides.

Everything must be predicated on diversity.

To take these imperatives into account and to reduce population growth will be a nearly impossible task. If there were any alternative, this overwhelming procedure would probably never be attempted. But there is no acceptable alternative. The alternative is the doomsday that sounds so feverishly unreal, and yet is inevitable unless new policies are adopted. This, it must always be remembered, is a finite earth.

Men individually will have to decide. Will the desire for another child come before the welfare of all? Will "progress" and the often rich promises of technology be sacrificed for less and for quality? Will a marsh be valued above a highway; a great lake more than an airport; an open, "unused" piece of land before a housing development?

Will a forest be saved because its diversity is a resource for the present and future? Will a manufacturer pay the cost of reducing pollution if it cuts seriously into profits, and can he?

Will a peregrine falcon, or the mud on the bottom of a pond, or the invisible plankton floating near the surface of the ocean be protected because they are part of the life-support system of the globe? Will the simplicity and efficiency of a single crop be given up for the diversity of planting that would insure stability? Will man recognize that he is a part of nature and not set apart from it?

René Dubos told us: "We can change our ways only if we develop a new social ethic—almost a new social religion. Whatever form this religion takes it will have to be based on harmony with nature as well as man, instead of the drive for mastery."

To make this decision, and to take the necessary steps to save ourselves and our environment, will first require the facing of serious philosophical questions by every man, woman, and child. We cannot, as Lynton K. Caldwell of Indiana University said,

> have a special ethic for the environment and another ethic for the rest of life. . . . If one were to name the single greatest cause of the ecological crisis of modern society, it would surely be this: that man has failed to unite science and ethics in a manner adequate to guide, restrain, and control individual and collective behavior in relation to the real world.[3]

The general ethics of the past may no longer guide us in a world transformed by science, technology, and the multiplication of man. The ethic of the Book of Genesis was an understandable consequence of our struggle to survive in a harsh and often unpredictable environment: nature then had to be subdued if man was to live. Later, when survival was assured, we sought to turn nature to our advantage. Nature was a mine from which we took the resources we could turn to our advantage for food, clothing, and shelter. Before the technoscientific age, there was little that man could do in this process to damage or irreversibly affect nature as a whole. But in making nature serve our material needs—the economic viewpoint—material wealth became our natural and proper goal. The mastery over nature, as Caldwell observed, seemed to become a fulfillment of human destiny.

Decisions we make—as heads of governments, as man-

---

[3] Lynton K. Caldwell, *Environment: A Challenge for Modern Society* (Published for the American Museum of Natural History; Garden City, N.Y.: The Natural History Press, 1970), p. 247.

agers of industry, or as individual consumers—have been made in this self-centered framework. In this day of a crowded, technically linked world, they have been made as though the actions of one in polluting the environment did not matter to others, as though we were still fighting for survival in a hostile wilderness, as though the mine to which we were going —the earth's resources—was inexhaustible, as though technology in its turn could conquer all, and as though human numbers were few.

In this modern world, in contrast, we will have to recognize that we cannot escape interaction with the total environment. We must learn that pollution poured into the air in England may have adverse effects on fishermen in Norway and that wastes dumped in the local creek may eventually pollute the seas. The environment will have to be understood as "something more than the sum total of its parts, more than aggregation of natural resources." [4] It will have to be understood as the basic resource of life and all man's hopes. And the environment, if this hope is to be realized, must include not only the physical land, air, and water, but the social, psychological, and ethical factors that are a part of it.

"To see the environment as a resource to be understood, conserved and utilized with regard to the total needs of the total man has become an ethical necessity," said Caldwell. ". . . Eternally and absolutely limited by the environment necessary to sustain his life, man's highly touted mastery of nature is a self deluding myth (and ethic)."

The new ethic, which will have to come first, before changes are made, may be discovered in the ideal of brotherhood so characteristic of the great religions of the world, and in a broadened Golden Rule. For now that we are becoming gradually and painfully aware of the extent of our dependence on the entire living world—for food, for oxygen, for the continued existence of a habitable environment—the Golden

---

[4] Caldwell, *Environment*, p. 156.

Rule could be extended to cover all of life. This is not a new extension of the ethic. It is deeply rooted in Buddhism, which long has held that kindness to any living thing is an expression of honor to all of life.

Life, though, is also dependent on the inorganic world, on the sun, the winds, the rains, the ocean currents, the composition of the rocks. Should the Golden Rule not be extended even further? Is it not true now that if man is to survive and live with dignity and grace he must enlarge his ethical horizons so that, in the words of Aldous Huxley, "the Golden Rule applies not only to the dealings of human individuals and human societies with one another, but also to their dealings with other living creatures and the planet on which we are all traveling through space and time." [5]

[5] Aldous L. Huxley, *Literature and Science* (New York: Harper & Row, Publishers, 1963), p. 109.

Chapter 7

IMPLICATIONS
OF CHANGE

It does not have to happen. Even now we have at least a part of the knowledge and technique necessary to halt a march to doomsday. But knowing what to do and wanting to do it will not be enough. To prevent overcrowding and the probably fatal upsetting of the natural systems that sustain life and the earth will require action, and in many cases early and expensive action. The financial costs of remaking the world as it is today could be astronomical, and the social and political costs—in the sense of altering many prerogatives—could be even higher.

No one at this stage can have the full answers. However, systems dynamics and other new studies of the earth's predicament are indicating what action may be required, some of the possibilities, and the implications.

The costs and promise of whatever is proposed will have to be faced and assessed. Action cannot be unprepared or blind. Only with the fullest possible understanding of the consequences can decisions be made and courses set. An at-

tempt will be made here to bring together the best informa-
tion developed so far on what can be done to cope with the
ecological imperatives, and on what can be expected.

A first necessity in moving into what might be called the
"fourth world"—if the ancient, the preindustrial, and the
industrial are considered the first three—will be a capability of
predicting the effects of any actions. Up to the present this
has been almost an impossibility. Both policy makers, whether
governmental or private, and voters have had to make their
own, largely unaided appraisals of what results might be. Each
formed his own mental image, or model, of what he hoped or
feared would be accomplished by any policy. Decision making
generally was guided by some interpretation of the past and
some vision of the future.

Sometimes the action taken achieved its objective. Almost as often it proved ineffective, or brought about exactly
the opposite result. In one case, foreign specialists acting with
the best and most unselfish of motives assisted the people in
the dry areas to the south of the Sahara in drilling wells. The
water enabled the people to expand their agriculture. At
about the same time, and as part of the same aid program,
vaccination and the worming of the cattle cut the former
grievous loss of animals. The result was that cattle herds increased at the exact time that agriculture was taking some of
the former range, leading to serious overgrazing. When prolonged drought came again, as it does to this drought-prone
area, disaster struck. The catastrophe devastated an area of
865,000 square miles in a belt of six countries south of the
Sahara, and the United Nations Food and Agriculture Organization told the world that more than 5 million people would
starve unless food was supplied.

Help was given under the direction of the United Nations, but the outcome is uncertain. A later study has warned
that the least disturbance can throw out of balance "the pre-

carious equilibrium between the soil, the plant cover, animals, and men." The desert may prevail; it has rolled south before.

Other instances of policy and action misfiring are nearly endless—the pollution of waters by fertilizer applied to the fields; air pollution from busy factories; the city-destroying effect of superhighways. The implication that must be faced is that man has seldom been able to foresee the full consequences of his actions.

Willis W. Harman, director of the Educational Policy Research Center of the Stanford Research Institute, argues that our inability to predict consequences is even more profound. We have not understood that the very successes of our technological era would inevitably create its problems. Harman cites some of the main examples:

1. Success in prolonging the life span has led to overpopulation and the problems of the aged.

2. The machine's replacement of manual and routine labor has exacerbated unemployment.

3. Efficiency in production has been accompanied by dehumanization.

4. Satisfaction of most of the basic needs of those living in the highly developed countries has led to rising expectations everywhere.

Jay W. Forrester, professor of management at the Massachusetts Institute of Technology, adds that some of our misjudgment and miscalculation about the effects of our policies must also be attributed to certain inadequacies in the human mind and experience. Both are ill adapted to deal with highly complex problems and the ramified interactions of the social and natural systems.

Said Forrester:

Until we come to a much better understanding of social systems, we should expect that attempts to develop corrective programs will continue to disappoint us.

The human mind is not adapted to interpreting . . . how

multi-loop nonlinear feedback [social and natural] systems work. In the long history of evolution it has not been necessary for man to understand these systems until recent historical times. Evolutionary processes have not given us the mental skill needed to properly interpret the dynamic behavior of the systems of which we have now become a part.[1]

The human mind, however, is quite adept in grasping the local consequences of some action, or the immediate consequences. The problem arises on the larger scale when one result interacts with another, and another, and still others over a long period of time. There failure occurs. We seldom can think our way through such webs, or if the technical word is used, model them.

The mind is also good at analyzing what is happening to us; what is paining and troubling us. Thus we can understand crowding, pollution, and depletion, and with some help can grasp the dangers of breakdowns in the natural chains and in the diversity of the natural world. We know what the trouble is.

We also are quite capable of setting goals. We know what we would like, and we can conceive or even visualize the ideal. We can develop theory.

As we come to the critical present, when the whole of nature and the whole population of the earth are affected by our actions, and when the human mind is not well equipped to follow through the dynamics and intricacies that result, a new development holds promise of supplying exactly the kind of expertise—predictability—needed. This is systems dynamics, a frontier pioneered particularly by Forrester and the associated *Limits to Growth* group. Its tool is the computer. As Forrester wrote:

It is now possible to take hypotheses about the separate parts of a system, . . . to combine them in a computer model and to

[1] "Counterintuitive Behavior of Social Systems," *Technology Review*, Vol. 73, No. 3.

learn the consequences. . . . The great uncertainty with mental models is the inability to anticipate the consequences of interactions between parts of the system. This uncertainty is totally eliminated in computer models. Given a stated set of assumptions, the computer traces the resulting consequences without doubt or error.

A number of other outstanding scientists agree that systems studies and the computer are giving man a new means for dealing with the overwhelming complexities of the present and future world.

Glenn Seaborg, 1951 Nobel Prize winner in chemistry and former chairman of the Atomic Energy Commission, quotes the British cybernetics expert Professor Stafford Beer as saying, "Society has become a complex organism and it needs a nervous system."

Seaborg continues:

We are a global civilization. . . . Today the computer is the vital link in that system. And in addition to telling us where we are, it can help us shape our future by giving us the means to project and examine the alternate futures. Through the computer models we can "look ahead" to the consequences of various courses of action we may choose. And thus we may choose more wisely.

Forrester is convinced that this—the development of understandings of the world systems—will be the next great era in human pioneering, much as geographical exploration was in the fifteenth and sixteenth centuries and science and technology in the nineteenth and twentieth.

Although there is strong disagreement about systems dynamics, most of the objectors concede that a computer model may more accurately forecast alternate results than the usual mental model—or thinking—of a policy maker. The promise is that the modern world will not have to move as blindly and gropingly as in the past. Reasonably accurate appraisals should be available.

So far only a beginning has been made, and only the earliest studies of the world's predicament have been initiated. Even at this point, however, the new studies have confirmed what the world has really known since the time of Malthus and Darwin—and perhaps earlier—that exponential expansion is not possible on a single planet. And this already points inevitably to another conclusion—that there will have to be change and cutbacks or restraint if life on this earth is to be sustainable.

To keep the earth viable, the systems analysts have already outlined one explicit course of action—simultaneously to limit population, industrial expansion, pollution, and depletion while increasing food production as much as possible. Their projections show that nothing less is likely to work. What would this require? What difficulties would have to be overcome? What are the implications? What would have to be done?

POPULATION

As a first step toward checking the present rampant growth of population, the establishment of an international control agency is proposed. In a world of separate independent nations, it is acknowledged that the authority of such an agency would be strictly limited. Education and persuasion would be the principal tools. The task of reducing birthrates would be complicated, too, by the racial and have or have-not divisions of the world. In the underdeveloped areas, where the populations are soaring, stricter birth limitation would be required than in the developed areas, where the birthrates—if not yet the numbers of people—are approaching the replacement level. The disparity in treatment would not be viewed with equanimity in the underdeveloped nations.

Populations are also growing rapidly in black and Oriental areas. A drive for more birth control in the nonwhite than in the generally white nations would also be regarded with the deepest suspicion. Charges have already been heard that birth control is an attempt to protect the white minority.

Only as the peoples of the earth realize that their interests and stake in this one world are the same will there be a chance of checking a disastrous growth of population. Such a change in basic beliefs would certainly depend upon a fairer sharing and a more equal protection of the earth's resources— a fact that will have to be faced sooner rather than later if populations are to be stabilized in time.

INDUSTRIAL GROWTH

The warning that industrial growth must be halted in a decade or two has been met with strident disagreement. The protests reveal what limitation would mean and cost.

For more than a century industrial growth has been an article of faith, a measure of success, the provider of goods, and also the solver of many economic problems. In earlier days in the United States it also took care of depletion. If a farmer exhausted his soil and no longer prospered, he moved on to newer, richer lands. In the industrial years, growth with its provision of jobs and creation of wealth was readily forgiven any ravishing of the environment. An annual growth rate of 7 percent was believed necessary to maintain the American standard of living.

The affluence that constant growth produced in the industrial countries was envied and sought by the other states. Poor countries look to industrialism as the best hope for overcoming their poverty and other ills, and until recently growth per se was all but unquestioned.

An ambassador from India, after listening to industrial growth limitation proposals, exclaimed: "Unless income were

equalized between wealthy and poor nations, the poor nations would slide down to starvation, while the wealthy would
continue to sap their resources."

Others charge the industrial nations with selfishness in
proposing or even considering limits at the very time the
poorer countries are trying to industrialize to lift their abys-
mally low standard of living. If technology were to be with-
held from them "for their own good," they are extremely
skeptical that the developed world would provide them with
the goods they want. Pollution seems a lesser evil than poverty,
or as Seaborg has said, "to the hungry and poverty-stricken
ecology is irrelevant." Several countries have already notified
industrialists that there will be no interference or pollution
standards if they will move factories to their lands.

The same problems would afflict international proposals
to restrict the use of fertilizers and limit the development of
vulnerable single-crop agriculture. Both are another phase of
growth. Again, efforts to prevent the abuses of fertilizers and
pesticides and to maintain diversity in agriculture would
seem to put the rich in the position of witholding bread from
the starving. The starving will not meekly agree unless they
are sure of the way out.

The underdeveloped countries also react negatively to
the idea that they should forgo relief for the hungry and needy
of today to protect their resources and land for the future.
Today is what matters. No eventual catastrophe is believed to
justify deprivation in the present, and few political leaders
in Asia, Africa, or South America would dare to suggest any
lessening of production to avert a disaster half a century or
more away.

In considering a limitation of industrial growth, account
must also be taken of one of the anomalies of population
growth. Even though birthrates should be reduced in the
nonindustrial parts of the world, those already born will produce additional millions, and populations may still double

by the end of the century. A huge increase in food, clothing, and shelter will be essential to support the larger population, and raising the standard of living would take much more.

The underdeveloped countries also believe that they must have industrial expansion to increase their wealth. During the past twenty years countries with a food deficit were able, if they could afford it, to buy grain at reasonable, stable prices from the principal grain-exporting countries. Those without funds could obtain what were called "concessional" sales or gifts from the United States. However, in the inflation of the 1970s the price of grain more than doubled, and the financial strain on purchasers increased. The countries aided by gift grain also knew that the United States Department of Agriculture had forecast that the concessional food demand would eventually exceed what the United States had to give away. Thus, more funds and the industrialization expected to supply them began to seem all the more vital.

To stabilize industrial growth, as well as population, at any time in the near or middle future will almost certainly require a radical new kind of sharing of the world's industrial wealth. At present the gap between the have and have-not countries is widening. It would be even less tolerable if the have-not countries were asked to refrain from any industrial development possible for them. On the other hand, some increases in plant would be permissible under any world plan—*The Limits to Growth* proposes 1985 as the checkpoint —and any expansion that did not cause pollution or a dangerous drain on resources could be encouraged and assisted.

Such conflicts in the interests of industrial and nonindustrial countries and white and nonwhite regions will be among the most difficult problems that will have to be solved if the world is to limit exponential industrial expansion. A whole new international approach will probably be necessary.

The big problem of industrial growth, however, is in the

already industrialized countries. To many in these areas even questioning future unlimited expansion is "mad," "unimaginable," and an unthinkable heresy. Others grant the dangers of exponential growth, but believe that it cannot be halted, or that there are alternative ways of dealing with the excesses of growth.

Three main arguments support the claim that unlimited growth cannot be stopped:

1. That there is no other possible way of supplying the needs of this country and the world—that only growing industry can produce the necessary volume of goods.

2. That only by growth can the conflicts between different economic groups be mitigated. Only if the pie is made larger can all have a larger share. If the pie remains the same size, the poorer can obtain more only by taking from the richer, and that would portend bitter struggle.

3. That there is no adequate machinery for bringing about the proposed stationary state, and that no Western statesman, any more than a leader of an underdeveloped country, could easily propose doing without in the present to protect the future. Could the social transition from exponential growth to equilibrium be made? These are charges and objections that will have to be met before limitation can be effected. Those who think there must be limitation are convinced they can be met.

In the first place, the level of production proposed in the first studies of the earth's predicament is a high one. With production now growing at 7 percent a year, the proposed 1985 ceiling level would be considerably higher than today's. If population growth should slow down, it should be adequate, or more than adequate.

In the second place, production in the affluent society of the United States contains much slack; it is widely conceded to be wasteful and unduly luxurious in many aspects. Commoner has also demonstrated that the substitution of synthetics for natural materials and fabrics has unduly increased pollution. And while a shift back to natural products would disrupt large and powerful industries, it would upbuild others. Such shifts have been made before.

If industrial growth were stabilized, capital growth would also be checked. Fortunes would be more difficult to accumulate, and some incomes might be lower, thus enforcing a less lavish mode of living. *The Limits to Growth* calculated that incomes in the United States might then more nearly approximate those of Europe. Many maintain that this would be a satisfactory level, and that none should have to suffer. The pie should be large enough for all.

The advocates of limitation also believe that American government and institutions could handle the transition to an equilibrium society, though changes and adaptations might be necessary.

POLLUTION

To reduce pollution by the 75 percent believed necessary might drive some industries out of business or require public subsidy to keep them going. Indeed some have already closed because the cost of "cleaning up" is excessive. Such closings may cause severe hardship to the owners, employees, and the cities dependent on them. The answer in such cases is public compensation and the construction of new plants that meet ecology standards.

Better handling of wastes and the recycling necessary to the maintenance of the earth will be costly and full of problems. A small start that has been made shows the difficulties and benefits that may be expected. In 1973 General Motors Corporation announced that it would build an $8.2 million plant to turn its own trash into fuel for heating and cooling one of its plants. Some 55,000 tons of scrap wood, cardboard, and paper are expected to produce 800 million pounds of steam. To produce the same volume of steam in conventional boilers would require 37,000 tons of coal—of course a nonrenewable fossil fuel.

The wastes will be trucked to a mill, where a shredder will cut them into small pieces, and a magnet and separator will draw off all unburnable elements, such as metal, stone, glass, or concrete. Not only will the usual pollution-creating disposal of trash be ended, but the company says that the substitution of the trash residue for the former high-sulfur-content coal will cut sulfur emissions from its stacks by 40 percent. Similar plants have been in use in Europe for more than thirty years, but the GM plant will be the first large industrial installation in the United States.

Others are expected to follow. In 1973 the city of Chicago announced its own plans for a $14 million plant to turn household refuse into a fuel for the use of the local electric utility. The company, Commonwealth Edison, has contracted to pay the city $3.60 a ton for burnable waste. The plant will process 250,000 tons a year.

Chicago expects to cut its waste disposal costs from $11 to $5 a ton; and the area will be spared the pollution and health menace of unsightly open dumps. The Chicago plant will also separate any metals or other hard materials in the refuse. After processing, the city anticipates making about $200,000 a year on the sale of the metals. A market will also be sought for the other separated materials, including plastics. The way is to be opened for the recycling of all of these materials.

A rapid spread of such plants may be expected, particularly if such handling of wastes proves to be economical. In this case, economy may be on the side of ecology.

A proposal has also been made, by the U.S. Army Corps of Engineers in the course of a waste-water study authorized by Congress, for pumping Chicago's waste water, including sewage and industrial wastes, into vast rural lagoons. Bacteria in the aerated ponds would be expected to consume the wastes. At regular intervals bottom residues would be collected for use as organic fertilizers, and the cleaned, chlorinated water would be sprinkled over croplands. The soil would serve as the final "living filter." The diversion of the wastes also would clean up the heavily polluted lower end of Lake Michigan and the rivers of the area. If the system is adopted and works as planned, pollution would be checked and the essential earth cycle completed.

Strenuous objections have arisen, however, in areas where the lagoons would be built. About 77,000 acres would be needed for lagoons large enough to handle the wastes of a city of 3.5 million, and another 365,000 acres for the sprinkled croplands. Development or heavy construction would be barred on another 500,000 adjoining acres. At public hearings in the Indiana areas that would be purchased or leased, some farmers assailed the plan as one to "dump the city's filth on us." Others objected that the 134 inches of water to be sprinkled on the fields would be excessive. They also resented the controls that would have to be part of the operation.

Chicago is already barging tons of sludge, prepared from sewage, to fields in Illinois. Some of the land on which it is being spread was ruined by strip mining, and the application of the rich organic material is restoring it to its natural productivity. But here, too, the community is opposing the operation. There has been little open recognition that the nutrients being returned to the soil could restore the natural life-giving cycles.

If the land is not to be destroyed by the constant withdrawal of nutrients, and if the waters are not to be dangerously polluted by these misdirected materials and by the outwash of synthetic fertilizers profligately applied to save the depleted fields, recycling projects will have to become the rule in all urban areas and the rural areas serving them—which will mean in a large part of the United States and other urbanized countries. This will probably entail stricter controls over the use of land than any known in the past. In some cases, a man's land will not be his to do with as he chooses.

The first systems developed to restore the earth's cycles and solve the formidable waste problems of cities may be far from those ultimately designed. The technology, however, is not difficult. Unlike fusion, it is well within the range of what is already known. The costs would be high—the army engineers have talked about $7 billion for the Chicago lagoon plan. But the stakes would also be high—the saving of the earth's cycles and its waters. Both will have to be done, and one fortunately abets the other.

The objections of the townspeople to the sludge and of the farmers to the lagoons are forerunners of the human obstacles that will be encountered in such undertakings, particularly in their early and undoubtedly imperfect stages. Both the stick and the carrot will be necessary—the adoption of standards to force recycling and money to perfect the processes and compensate any who may be damaged. New understanding of the ecological necessity of such action will also be needed—and that will mean more ecological education.

NATURAL RESOURCES

If population and industrial growth were stabilized, the drain on natural resources would be substantially relieved and the use of nonrenewable resources would be reduced. If this were not enough, and surveys have suggested that the use of certain critical resources would have to be reduced much more, additional restrictions would have to be imposed. It has already been suggested that the use of natural gas should be prohibited in processes for which other fuels could be burned. The country also could expect a decrease in "throw-aways" made of scarce natural resources. Longer-lasting products would be demanded, and with this, yearly model changes might be abolished. The standards would be applied to the whole range of machinery, appliances, and products made of metal and other limited resources or of pollution-producing plastics.

The achievement of equilibrium would not necessarily protect the essential diversity of nature. New, diversified methods of planting would have to be developed and sold to the farmers of the world—as other new systems have been in the past. Public action, nationally and internationally, would undoubtedly be required to protect forests, shores, marshes, perhaps even mud, and the present diversity of species.

Struggles would be unavoidable and might take many directions. In California one conservation group undertook and won a fight to stop the intrusion of a large resort into a unique wilderness. California also adopted, with 55 percent of the vote, an initiative petition to control further commercial development along the thousand-mile state coastline.

In time international bodies might be forced into action to save tropical jungles and rain forests or other special areas, such as the northern tundra. In either a growth world or an equilibrium state, the struggle to save diversity would be a fierce and continuing one.

Even the foreseeable costs, social and economic, of turning to an equilibrium state are of a magnitude never previously encountered. The costs of the past were those of growth, and therefore easier to absorb. Without exponential or unlimited growth the people of the United States and other industrialized countries would have to be prepared to accept a less wasteful, less superaffluent, less expensive way of life. Institutions and even the way of thought might have to be changed.

Nevertheless, the cost of equilibrium does not appear impossible. And exponential growth does not seem to be the only way of satisfying world requirements until it leads to catastrophe.

The Meadowses and others studying this greatest problem of them all—actually the survival of the earth as a habitable planet—find that equilibrium could supply a good and ample life for all. They maintain that there could be more food, more services of every kind, more leisure, a livable environment, and perhaps a more even, less harried life. Even the industrialist, if he were not driven to increase profits each year but only to maintain and improve the quality of his product, might discover a better, if different, life.

The Meadowses argue:

Historically mankind's longest record of new inventions has resulted in crowding, deterioration of the environment and greater social inequality because greater productivity has been absorbed by population and capital growth. There is no reason why higher productivity could not be translated into a higher standard of living or more leisure, or more pleasant surroundings for everyone, if these goals replace growth as the primary value of society.[2]

The computer scientists were not the first to cite the values of an equilibrium society. More than a hundred years ago John Stuart Mill wrote:

I cannot . . . regard the stationary state of capital and wealth with the unaffected aversion so generally manifested towards it by political economists of the old school. . . .

It is scarcely necessary to remark that a stationary condition of capital and population implies no stationary state of human improvement. There would be as much scope as ever for all kinds of mental culture and moral and social progress; as much room for improving the Art of Living and much more likelihood of its being improved, when minds cease to be engrossed by the art of getting on.[3]

2 *The Limits to Growth*, pp. 178–79.

3 *Principles of a Political Economy*, in *The Collected Works of John Stuart Mill* (Toronto: University of Toronto Press, 1965), p. 754.

The contention is therefore that a limited, nonexpanding industrialism could supply the needs of a limited, non-expanding population, if new and possibly better, though less affluent, standards were accepted. The key words are "limited" growth, "limited" population, and "acceptance" of a different standard of living.

Other students of the country's and the world's predicament emphasize that in the "real" world the increase in the number of people cannot possibly be halted before at least the first decade of the twenty-first century. They insist that only a growing industrialism can feed, clothe, and shelter the 400 million expected in the United States by the year 2000 or the 7 billion of the world.

Robert L. Heilbroner, professor at the New School for Social Research, pointed up the implications:

The fact of a relentless burgeoning of population—and the hope for a rise in their material consumption—makes inevitable the need for very large increases in physical output. Huge additions to food production, textile output and simple shelters will be required to sustain, much less elevate, the prospective billions of the backward world. In turn this requires the output of vast quantities of fertilizer, of steel, of cement, and bricks and lumber, with all the environmental problems. That vast—although indeterminate—increase in needed output provides a powerful incentive to rethink the desirability of "zero industrial growth." [4]

Heilbroner predicted that, when the insistent growth demands of the industrialized world are added to those of the underdeveloped countries, it will be extremely difficult to stop increases in production, and that those increases will not be stopped if they are attainable.

On the assumption that industrial growth will have to continue for a much longer period than the MIT studies indicate that it should, Heilbroner examined the possibilities of preventing the worst damages:

[4] "Growth and Survival," *Foreign Affairs*, Vol. 51, No. 1, p. 145. All other quotations by Heilbroner in this chapter are taken from this article.

It is not growth that is the mortal enemy, but pollution. Any technological change that will increase output without further damaging the air or water or soil and any technological change that will enable us to increase output by shifting from a less to a more abundant resource (again without any increase in pollution) represents perfectly safe growth and should be welcomed with open arms.

This appears to be very close to the Meadows–*Limits to Growth* position—that any activity that does not damage the environment or draw too heavily on nonrenewable resources could expand indefinitely. There are, however, differences in the two approaches.

If production must be increased in a large way, would the consequences be the exhaustion of resources and increase in pollution that is predicted in *The Limits to Growth?*

Despite estimates that nonrenewable resources, particularly metals and oil, cannot last long if usage continues to increase, Heilbroner emphasizes that "the full extent of resources" is not known. "Proved" reserves have repeatedly been increased—despite heavy usage—and "known" reserves have seemingly been dependent upon how vigorous a search has been made for them. Little is known of what may lie under the frozen wastes of Siberia and the pole areas, and exploration of the sea floor is also in the earliest of stages.

The Limits to Growth demonstrated that exhaustion would come even if exploration and discovery should multiply supplies of nonrenewables by five. Heilbroner suggests, however, that actual resources might prove to be ten to fifty times greater than the present estimates. In such a case the day of exhaustion would be delayed. It could not, of course, be averted forever.

Most of the figures on how long supplies will last consider only the separate minerals or other resources. Little attention is given to what is called "substitutability." Heilbroner argues that if the reserves of oil and gas are exhausted

in another generation, as now seems likely, the enormous
stores of oil shales might be drawn upon. He points out that
when the best iron ore of the Mesabi Range was depleted,
lower grades of ore were mined. He concludes therefore that
resource substitution may also defer the day of reckoning
for an indeterminate time. In Heilbroner's judgment the
pushing back of the present resource limit is all important,
for he believes that potentially the earth has boundless resources "locked in the rocks and sea waters." If, he maintains, the necessary technology were developed and sufficient
power were supplied, probably from nuclear sources, the
rocks could be melted and almost any resource reconstructed
by synthetic processes. He continues, in "Growth and Survival":

To be sure, such processes would entail the processing of
enormous quantities of sea water or granite, with associated
problems of disposal and thermal pollution. But from the point
of view of sheer bottlenecks of supply, the long-term future holds
out much more promise than the anti-growth schools of thought
reveal.

The key problem in this analysis is conceded to be
time—time to discover the technology. If it should appear
that this period was going to be lengthy, Heilbroner would
advocate the strictest husbanding and recycling of scarce resources.

"Thus," he says, "the resource problem hinges finally on
our scientific and technological capabilities. Judging from the
past it would be foolish to take an attitude of determined
pessimism before these capabilities . . . the exponential curve
of that resource soars steadily upward without sign of limit."

Other critics have pointed out that *The Limits to
Growth* does not anticipate an exponential growth of technology, though such growth has occurred during most of the
scientific age and is seemingly occurring today. To illustrate

the effects, some cite the telephone. If it were restricted to the technology of 1900, some 20 million operators would be needed to handle the volume of phone calls today. (One British editor wryly observed that an extrapolation of the trends of the 1880s would have today's cities buried under horse manure.) This group scoffs at the idea, as advanced in *The Limits to Growth,* that technical progress cannot keep pace with needs, that exploration and recycling can do no more than double capacity, that agricultural research can only double yields, and that pollution cannot be reduced more than three-fourths. Faith is placed in technology to out-run and solve all problems.

Some of the technologists, or advocates of technology, admit the difficulties of their recommended course. To produce the power and raw materials for a growing industry would almost certainly produce a phenomenal lot of pollution. *The Limits to Growth* assumes that 75 percent of present pollution must be eliminated.

Heilbroner, although he is convinced that economic growth must continue and that some of the worst industrial pollution can be mitigated, is nevertheless concerned about two types of pollution. The first is that produced by combustion. Without any question, the burning of coal is increasing the amount of carbon dioxide in the atmosphere. Since CO_2 makes up only about 1 percent of the amosphere—it is called a "trace gas"—relatively small additions to the supply can have a large effect. If expanding industry should increase the CO_2 content of the air by one-third, there might be change in the air's absorption of infrared radiation and its reradiating of this heat back to earth. Temperatures might increase—and only a few more degrees could melt the polar ice caps. If that happened, the coastal cities of the world from New York to Hong Kong could be covered by water. The oceans would roll over them.

A shift away from the combustion of fossil fuels and the development of nuclear and hydrogen power would, however, solve the problem of carbon dioxide. The construction of nuclear power plants is already well advanced, and by the 1980s some of the large cities of the country expect to obtain nearly half of their power from nuclear sources. Studies of fusion, or the production of power from hydrogen, are also promising, although some of the basic problems have not yet been solved, and no commercial extraction of power from seawater is in sight in the immediate future.

The use of nuclear power, however, would not avoid the heat problem and other problems of pollution. Nuclear plants have heated entire lakes and rivers, with consequences that have not yet been fully appraised. The disposal of nuclear wastes is also proving so difficult and dangerous that some environmentalists have warned against entering a "nuclear age," and have urged the country to forgo it.

Could the heat pollution problem be overcome if industry continues to grow at the present exponential rate? What would this entail? Heilbroner shows that no accurate measures have been made of man's heat pollution and its effects. Nor have comparisons been made with the heat produced by volcanoes, hot springs, and the steady heat input of the sun. Heilbroner also suggests that the technology of coping with heat pollution might grow as exponentially as heat pollution proper.

A third school stands in between those who insist growth must be limited and those who take the position that technology will find a way to permit capital growth to continue for a considerable time in the future. Seaborg is one who thinks that the world is not now faced with an either/or proposition. He and others of this school say that there are areas where new growth and development are essential and others in which they should be leveled off or cut back.

"The key to the middle way," said Seaborg in a symposium, "Future Realities," "lies mainly in the wise development and application of science and technology."

He continued:

There is no way to reduce poverty without greater productivity and economic growth. And these are tied to advances in science and technology. There is no way to reduce pollution and solve our environmental crisis without greater operating efficiency and the re-cycle of resources. And these are tied to advances in science and technology. There is no way to reduce the growth of population without a combination of better methods of birth control, better education, and a better standard of living for all. And these are tied to advances in science and technology.[5]

Another advocate of a middle course suggests that it could be made to work by the use of the price system. Henry C. Wallich, economist of Yale University, argues for discouraging pollution and over-use of nonrenewable resources by making both costly. If polluters had to pay for the damage they do to the environment, and if consumers of natural resources could not afford wastefulness, Wallich predicts that the most injurious kind of industrial growth would be slowed.

"The price system can be made to function . . . as a sensor for an emergency brake that will stop growth when the costs become too high," he says.

A factory that had to invest heavily in pollution-reducing equipment, and that had to pay heavily for scarce raw materials and land, might soon not be making very much money, and the incentive to expand would thus be squelched. Output would also tend to shift toward low-pollution components.

Pollution also could be taxed directly, as well. By making the polluter pay for any damages done to the environment, pollution might be reduced directly as well as indirectly by checking expansion.

Wallich cites an imposing array of mechanisms to deal with resource shortages. If a resource, such as copper, became exceedingly scarce, the price quite certainly would rise. The higher price would activate the working of lower-grade ores, recycling, the substitution of other materials, and research. If all of this should fail, however, prices would probably rise so high that demand would fall.

Though prices rise when scarcity occurs, they do not always rise in anticipation of future shortages. Buyers reason that the shortage may never come to pass, or that it will be overcome by other developments before it becomes severe. Because of this chronic inability to face up to the future, Wallich would be willing to see a resource tax imposed to conserve potentially scarce materials. He would also favor taxation to check increases in the price of land.

If costs rose sharply and industrialists had to reflect the increases in the prices of their products, the price "brakes" would be activated. As Wallich wrote:

Investment would decline as its returns diminish. Growth would slow down and perhaps eventually come to a halt. The declining rate of return would reduce the income of owners of capital. The distribution of income would become much more even. This is an important part of the slowing down process.

The economist cautioned that the attainment of a stationary state in the United States would not solve the ecology predicament in the rest of the world. A worldwide cessation of economic growth would be necessary, but in this view it would have to rest on a cessation of population growth.

Wallich readily grants that in a finite world growth cannot continue indefinitely, but he fears "prematurely denying society the benefits of growth.

"It makes more sense," he said, "to accept the benefits

but adopt protective measures. If they work properly, un-desirable effects of growth will induce feedback that slows or halts the particular kinds of growth producing those ef-fects."

Strict and early control of growth . . . continuing growth with reliance on science and technology to handle the worst effects and carry the earth along . . . strict control only in certain critical areas, along with the use of science and tech-nology to solve urgent problems . . . or use of the price sys-tem as a brake on dangerous growth. . . . These are the main actions, or among the main actions, that the world will have to take if it is to avoid calamity and reduction by degenera-tion and collapse. It is the kind of action that will have to be taken to keep the earth functioning ecologically.

All agree that action will be necessary. The world can no longer let things go along as they always have. The dis-agreement is only over timing and emphasis, decisions that have not yet been made. Neither is it clear who will make them and how it will be done, except that the attitude and beliefs of the billions of the world will have to change.

It is highly likely, however, that if the transition is to be made, changes will be required in the organization of the world and in its political and economic systems. Certainly world action will be necessary to protect the world environ-ment and to regulate growth.

At the very least, this will call for the formation of many new, specific world organizations, and perhaps for a new over-all world organization, more effective than any in the past. Such organizations would set standards, sponsor research, monitor danger points, and if possible persuade the various countries to adapt their policies and practices. For such or-ganizations to succeed in a world where shortages and pollu-tion are threatening will be exceedingly difficult; interna-tional organization has not yet been able to stop war or abolish hunger. Will jealously guarded national interests be partly relinquished to save the world at large, or will each put the feeding and survival of its own people first?

Even if all goes as well as possible, world stress can be anticipated, because as society begins to overtake the natural capacity of the earth, strains will multiply. When shortages develop in any area, a first step will be to make purchases or seek aid from others. This will increase the dependence of one on another. However, even the largest producers and those with surpluses cannot indefinitely meet greatly enlarged demands without running against their own limits. International trade, another international institution, has always eased shortage by redistributing excesses, but under extreme demand it tends to break down. For this reason, the entire world will likely feel any pinch at about the same time.

How quickly strain may come was seen in 1973. When unfavorable weather in Russia and drought in the countries south of the Sahara forced those countries to seek wheat in the international market, prices soared and the wheat stocks held by the four major exporters dropped to the lowest point in twenty years. The world suddenly became aware of shortage.

The tendency in any case of stringency will be to relieve its pressures as long as possible. When that becomes impossible, the world will abruptly find itself confronted with new shortages—which will seem to develop overnight.

If shortage and pollution become severe, many states will find their political and economic systems under pressures of their own. Heilbroner questions whether capitalism can easily adjust to the measures that would be necessary—or to a stabilized economy—in which "growth no longer tempers the struggle over the division of the social product." Some national states might have to be cast into a different mold. But Heilbroner was not suggesting that the socialist and communist states would be any better off or would have any more of the answer. They, too, would be confronted with the same crisis and necessity for change.

"Given the constraints of ecological limits that cannot be safely breached," he said, "the ideological premise must be abandoned and the incentives of true communism rethought."

If nothing effective were done, it is doubtful that any existing form of government or economic system could long survive approaching starvation and the ruin of the earth. Whatever the difficulties of meeting the needs of a limited earth, with new organizations, the trauma of doing nothing would be infinitely worse.

There is only the certainty that if the earth's greatest crisis is to be surmounted, the attitude of men toward their worsening world will have to change. Some of the most profound and ingrained beliefs will have to be altered. And the change will have to be widespread, not just in a few parts or in one section of the world.

Only rarely in the history of modern man have such changes come about, changes in what Thomas S. Kuhn has called "the basic way of perceiving, thinking and doing, associated with a particular vision of reality." Kuhn added that this perception or view of reality must be transmitted, not so much formally as tacitly, through understandings and what he termed "examplars." It must almost be taken in like the air.

After the fall of Rome the world was perceived in a new context. The Reformation marked another division, and the industrial revolution brought another. After each of these changes, the people of the world began to see life and the earth and values far differently than had those who preceded them. Until recently in the Western world, the view that came with the industrial revolution dominated. Few questioned man's right to control nature, industrialism, the division of labor, growth, the scientific method, or materialism.

Exactly how another view might come about is certainly not known. There is evidence, however, particularly among the young, that a new attitude is developing. Many are beginning to recognize that man and nature are one; that public and world good, as well as private gain, must be a goal; that the future cannot be sacrificed for the gratification of the present—or in summary, as Huxley said, that the Golden Rule must apply for other forms of life and for the environment as well as for man.

Man has made such change, or changes equally profound, in the past. There is hope that these will be made again, and that man will learn how to go on living on the thin skin and in the thin film of atmosphere on this small, vulnerable planet. Man must learn. The alternative is too dreadful to permit.

Index

A NOTE ABOUT THE AUTHOR

Ruth Moore was born in St. Louis, Missouri, and received her B.A. and M.A. degrees from Washington University there. She worked as a reporter for newspapers in St. Louis, Washington, D.C., and Chicago, in the process becoming a well-known science writer. Her first book, Man, Time, and Fossils *(1953), has been translated into ten languages. Her biography of Charles Darwin was published in 1955,* The Earth We Live On *in 1956,* The Coil of Life *in 1961, and* Niels Bohr *in 1966. She is now president of the Women's Board at the University of Chicago, president of the Prairie Avenue Historic District, and a member of the Chicago Landmark Commission.*

A NOTE ON THE TYPE

The text of this book was set on the Linotype in a type face called Baskerville. The face is a facsimile reproduction of types cast from molds made for John Baskerville (1706–75) from his designs. The punches for the revived Linotype Baskerville were cut under the supervision of the English printer George W. Jones.

John Baskerville's original face was one of the forerunners of the type style known as "modern face" to printers—a "modern" of the period A.D. 1800.

The book was composed at American Book–Stratford Press, Inc., Saddle Brook, New Jersey. It was printed and bound at The Colonial Press, Clinton, Massachusetts. Typography and binding design by Gwen Townsend.